FLOORS, WALLS & CEILINGS

FLOORS, WALLS & CEILINGS

**Graham Blackburn
and the Editors of Consumer Reports Books**

Consumers Union
Mount Vernon, New York

Library of Congress Cataloging-in-Publication Data
Blackburn, Graham, 1940–
Floors, walls & ceilings.
Includes index.
1. Floors. 2. Walls. 3. Ceilings. I. Consumer
Reports Books. II. Title. III. Title: Floors, walls,
and ceilings.
TH2521.B57   1989   690'.16   88-71028
ISBN 0-89043-245-7

Design by Jeff Ward
Illustrations by Graham Blackburn
First printing, May 1989
Manufactured in the United States of America

Contents

INTRODUCTION

Few things can make a home look better than beautifully finished interior surfaces. They dramatically improve a home's appearance, liveability, and value. On the other hand, few things detract more from a home's interior than worn floors, dirty walls, and cracked ceilings. Whether you live in a small apartment or a large house, whether it's brand-new or a vintage Colonial, *Floors, Walls & Ceilings* can help you achieve high-quality results when repairing and refinishing those structural surfaces of your home and can help you keep your outlay of time, effort, and money to a minimum.

Floors, Walls & Ceilings is written for people who have a minimum of skills but who would rather do the job themselves. It provides detailed information about how to plan and what materials to use. Step-by-step instructions explain the best ways to prepare surfaces and the practical repair and installation techniques that should be used to get the job done correctly. With easy-to-follow illustrations, *Floors, Walls & Ceilings* is designed to help you use the right materials, tools, and techniques to get the best results.

HOW THIS BOOK IS ARRANGED

Any living space is a cohesive whole, and its parts are interrelated, but for convenience this book has been divided into three main sections: Floors, Walls, and Ceilings. In reality, you'll find it difficult to work on any one area without involving, in some way, at least one other. For example, if you install a new tile floor, you'll probably have to touch up or repaint the room's baseboard, and you may well want to repaint the room's ceiling, walls, and trim to complement the new floor. Before patching a cracked wall, you'll need to ascertain what type of wall it is and whether the supporting structure and surrounding surface are sound.

Before you attempt any particular project, it is important that you make yourself familiar with the structures underlying the surfaces and the techniques for repairing and refinishing the surfaces themselves: you'll find it easier to isolate problems and do the work in a logical order. Familiarity with the underlying structures helps prevent one project from creating two or more projects. For example, if you discover a structural ceiling problem *after* refinishing an entire room, you must repair the ceiling and then go back and redo your paint job. Nevertheless, within each category, the various projects have been kept as self-contained as possible. Once you have gained a basic knowledge of the underlying structures, you should be able to use the book as a reference just for those subjects that concern you.

Each section begins with an overview of the structures involved and the various types of finish treatments commonly found in houses and apartments. After a discussion of repairs, you'll find instructions for a more extensive renovation, including installation procedures.

WHAT IS COVERED

Through the years, the products and processes of residential construction have changed. Although certain aspects of construction have remained constant, many variations exist, some new and innovative, others obsolete and archaic. This book cannot pretend to be a guide to everything you may encounter. Areas that should best be left to experts are noted— for example, complicated plaster molding, ornate woodwork, and fancy metalwork. We also caution you about jobs that we feel are more difficult or that may require a qualified engineer's or contractor's expertise—for example, substantial foundation work and other questions about structural integrity. But with planning and care, you should be able to accomplish most jobs discussed in this book. Finishing and refinishing floors, walls, and ceilings is labor-intensive; it does require a commitment of time and effort to do the job well. If you're tempted to cut a corner or two, bear in mind that a job well done not only improves where you live and increases your home's value, but it can also give you a sense of accomplishment and the feeling that your home is more "yours."

SAFETY

There are two systems that should always be approached with extreme caution: electricity and plumbing. In some communities, only licensed tradespeople are permitted to do this kind of work. The book points out when you are most likely to encounter wiring or plumbing work or the necessity of dealing with those systems so you can avoid the problems or plan for outside help.

The fumes and vapors given off by adhesives, solvents, clear finishes, and paints—even water-based paints—should be taken seriously. Throughout the book we recommend you always ventilate a room well when working with these products: establish cross-ventilation with open doors and windows and a window fan set to exhaust air. Read labels carefully, observe all warnings. In addition, as a matter of prudence, expose yourself and your family to these fumes as little as possible. Keep children, older family members, and pets away from the area. Of course, keep these and other household chemicals out of the reach of children.

We occasionally give you recommendations about other safety equipment you should use when working with finishing products or tools—for example, protective goggles, masks, or clothing. We urge you to follow those precautions.

TOOLS AND PRODUCT RATINGS

At the end of the book, you'll find an appendix of selected product Ratings and a glossary of tools. Almost every basic tool mentioned in the text can be found in the glossary, with a fuller explanation of its use. "Use the right tool for the right job" is a reliable carpenter's adage, one we ascribe to. But since you are unlikely to attempt everything in the book, you needn't equip yourself with every tool listed. In general, buy tools only as you need them, and buy good-quality tools. Don't think you have to rush out and clear the hardware store's shelves to get started. In fact, it sometimes makes more sense to rent equipment; when appropriate, we point out that alternative in the planning section of a project.

Finally, "measure twice, cut once," another carpenter's adage, contains a full measure of worthwhile philosophy: plan ahead, take your time, work carefully. If you follow those prescriptions, *Floors, Walls & Ceilings* will provide you with the information that can make your effort well worth it.

PART 1

FLOORS

1

HOW FLOORS ARE CONSTRUCTED

Practically every floor in your house is either a solid-base floor or a suspended floor. A *solid-base floor* is constructed directly on the ground; it usually consists of a slab of poured concrete with its top surface finished in one of various ways. A *suspended floor* is one with space below it. This space may be as low as a 2-foot crawl area, or it may be a full-height basement or other room. In a modern high-rise apartment, a suspended floor may also be a poured slab, but by far the greater majority of suspended floors are constructed with a system of joists, subflooring, and finish floor.

Whether flooring is solid-base or suspended, its finish surface may be wood, vinyl, ceramic, stone, carpet, or other material. Which of these materials is used—and how it is laid—depends partly on which type of floor it covers. Once you look at how the two main types are constructed, you'll understand the kinds of problems that can affect different finish materials. The remedies and solutions detailed in the pages that follow will then make more sense.

SOLID-BASE FLOOR CONSTRUCTION

In those areas of the country where the temperature drops below freezing, a foundation built below the frost line is required. That usually results in houses with basements below ground. Thus, only the basement floor will be a solid-base floor; all other floors will be suspended. In areas where freezing does not occur, it is common for a concrete slab to be poured directly on the prepared ground (see Figure 1). The house is then constructed on the slab, dispensing with a basement or even a crawl space.

Whether the slab is poured within previously prepared foundation footings or into specially prepared forms, the construction is similar. The first step is to prepare the ground on which the slab will rest. In some areas a *load-bearing test* is required first by the building code, but in any case the contractor must make sure that the ground is level, firm, and well drained. Leveling and firming are done by a bulldozer; drainage is achieved by laying a bed of gravel or crushed stone; then finally, the concrete is poured into the defined area and leveled.

The process can be complicated by the inclusion of various other elements in the slab. A moisture barrier, for example, in the form of impermeable plastic sheeting, is usually positioned between the prepared base and the slab. This is designed to prevent the penetration of groundwater or moisture up through the surface of the slab. Plumbing lines, heating ducts, and wiring cables may also be set into the slab. Occasionally, heating elements themselves will be built right into it. Reinforcement of the concrete is most important, to reduce the possibility that the slab will crack or develop fissures later on. This can take the form of special reinforcing rods or sheets of wire mesh—you may have seen these used in a highway under construction.

The final leveling and smoothing of the slab can be done to differing degrees, depending on what type of finish floor, if any, is to be used. As a garage floor, for example, the slab may be left as poured; as a basement floor, it may be rendered smooth enough to paint. But however the slab is finished, the basic construction of it remains the same.

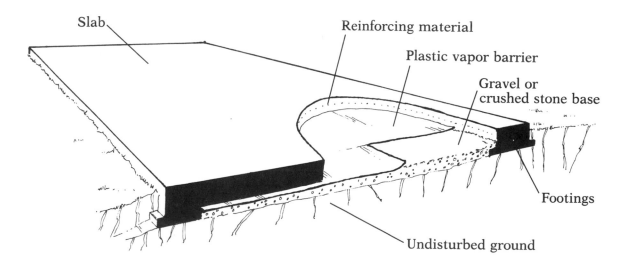

Figure 1   Solid-base floor construction

SUSPENDED-FLOOR CONSTRUCTION

A suspended floor includes a number of different elements (see Figure 2). The joist is perhaps the best known, but the other elements are equally important and must be considered if the whole system is to be properly understood. Since there are some differences between the first floor and the subsequent floors, we'll deal with the first floor first.

First-Floor Construction

On the first floor, the *joists*—the horizontal wood beams that support the flooring proper— are themselves supported at their ends by *sills,* which are boards resting on, and usually bolted to, the top of the foundation. If the floor is much wider than 10 or 12 feet, the joists will also be supported at their center by another beam, called a *girder,* which is set into the foundation wall at its ends. Like the joists, the girder, if it is long enough to require it, will be supported at various points by *piers* and *columns.*

If the distance a joist is required to span is great enough, it may actually consist of more than one board. In this case, the ends of the boards forming the joist overlap and together rest on a girder.

Joists vary in size—they can be two-by-sixes, two-by-eights, two-by-tens, to two-by-twelves. The actual size depends on the width they must span and is usually dictated by the local building code. The code's specification is based on the weight they will have to

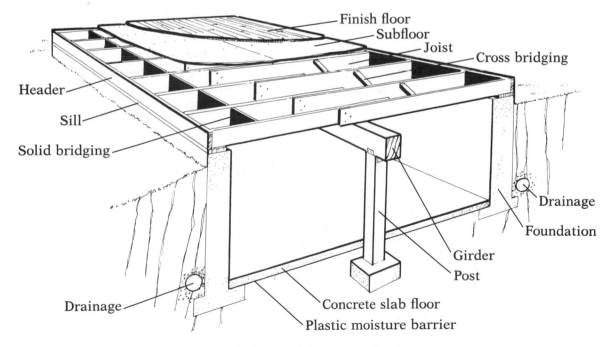

Figure 2 Suspended-floor construction

bear, assuming that they'll be set 16 inches apart, the current norm. In older buildings, however, it is not unusual to find joists set at different spacings. In such cases, the longer the distance beween joists, the heavier the joists will be. In old barns, for example, you may find ten-by-ten joists spaced as much as three feet apart.

Where the ends of the joists rest on the sills, they are attached to a finishing board called a *header*. Headers are also used to finish off any interruptions in joists, such as a stairwell opening. At various places in the framework of joists—at either side of openings in the floor, for example, or at places where interior walls will be built—two boards may be used spiked together to form a double joist for extra support.

Bridging

In good-quality construction, extra pieces called *bridging* are inserted between neighboring joists, primarily to help maintain the rigidity of the floor. It may also help support any walls erected on the floor at points not located immediately over a joist. Although *solid bridging*—material the same size as the joists—is often used, the more common material is smaller one-by-threes. These are installed in an X pattern, called *cross bridging,* in which one end is attached to the top of a joist, the other end attached to the bottom of the adjacent joist, with another piece attached in the reverse manner, right next to it. It is not uncommon to find bridging made of metal straps.

Girders

The girders that support the joists may be made of heavier wooden beams, such as eight-by-eights (4 two-by-eights spiked together), or they may be steel beams. Since houses come in all sizes and shapes, it is difficult to generalize, but the girders will usually run the length of the building rather than across its width. This is done to reduce the span of the joists. Keep this in mind if you're trying to locate the position and direction of concealed floor beams.

Subfloors

In the earliest form of floor construction, the boards laid on top of the joists were the actual finish floor. These were often just random-width boards, usually at least an inch thick, sometimes plain-edged but more often matched to fit into each other at their edges. Later, a two-part system consisting of a *subfloor*—rougher boards or lower-grade plywood—and a finish floor became the normal construction. Now, once again, it is not uncommon to find a single layer of flooring being used directly over joists, but the one layer is a more sophisticated material intended as *underlayment* (see page 33) for floor materials such as carpeting or tile.

Subflooring was originally made of boards of various widths, generally laid diagonally across the joists so that whichever way the finish wood floor was laid in relation to the joists—parallel with them or at right angles to them—it was still diagonal to the subfloor and consequently always found substantial nailing surface. To help eliminate squeaks and drafts, a layer of felt paper or rosin paper was frequently inserted between the subfloor and the finish floor. Plywood or other sheet forms have now commonly replaced boards for the subfloor, since they take far less time to install and allow the laying of the finish floor in any direction.

Upper-Floor Construction

The difference between first-floor and upper-floor construction lies primarily in the way the joists are supported. If the outer walls of the building are made of masonry, then the joists are supported at their ends at all subsequent levels just as they are at the first level—set into, or resting on ledges set into, the walls. If the building is of wood-framed construction, the usual method is to rest the ends of the joists on the *plates,* the horizontal boards topping off the frame wall of the floor below. The second-floor walls are then built on top of the second-floor joists, and so on, until the roof is reached. In some cases, the walls of the building are built in one piece to their full height; the ends of the second-floor joists are then tied into the wall framing wherever they happen to intersect.

Interior load-bearing walls built on the first floor take the place of girders for the second floor; in the same way, they serve to reduce the span of the joists. This system continues up through the subsequent floors of the house.

2

HARDWOOD FLOORS

ASSESSING THE CONDITION

Under the heading of hardwood floors come all types of finish floors that are made of wood—strip flooring, parquet flooring, and wide-board floorings of both soft and hard woods. Soft woods, such as pine, are not a good choice because of the wear that most floors have to withstand. Nevertheless, pine flooring is sometimes found in older houses or in houses that have been built with an eye to economy. Note, however, that in many of the older houses, the pine floors are more likely to have been made out of hard yellow pine rather than soft white pine. You can't really use color to tell the difference, since both kinds tend to turn to rich amber as they age. Soft pine is far more likely to show excessive wear and numerous marks from high heels, chairs, and table legs.

Before assessing the actual wood surface, you should check the floor's basic condition. Any repairing or refinishing of the floor's surface is likely to be wasted unless you deal first with problems in the underlying structure—sagging, excessive give or springing, or noises that can be traced to causes below the surface. We'll deal with each of those problems in turn.

Much of what follows is applicable to all suspended floors, no matter what finish material has been used. Keep that in mind when assessing the condition of floors described in subsequent chapters.

REPAIRS

There are several reasons why a floor may slope or sag. Although some problems may require professional help, not all of them are serious, and some of them do not warrant an attempt to rectify the problem.

Settling Problems

Settling is a common cause of uneven floors, but the settling may long since have stopped. If the house is more than about forty years old, it is probably quite stable, with little chance of further settling. Assuming you can live with the "character" a slightly sloping floor often imparts to an old farmhouse, for example, there is no reason to undertake complicated and expensive repairs.

To decide whether the settling occurred shortly after the house's construction and then ceased, or is an ongoing condition, look at the walls and ceilings and around doors and windows. Open cracks or doors and windows aslant in their frames and increasingly difficult to open or shut are signs that the settling continues. But if there is evidence that related problems with cracks, doors, or windows were repaired long ago, you can assume that the house is now stable.

If you suspect that the settling continues, try to find out why. Inherent structural weakness may be one reason. Older houses, especially those built with nonstandard framing techniques such as beamed ceilings and irregularly spaced wall framing, often sag over the years. Floors that sag because the joists or girders are too few or too far between can be leveled if the slope is intolerable, but the sag doesn't necessarily constitute danger. If there is any question about the cause of the settling, consult an engineer for a definitive opinion.

On the other hand, continued settling is sometimes the result of serious deterioration in the foundation or the flooring system. Possible causes are rot, insect damage (from powder post beetles, carpenter ants, or termites, for example), settling of the ground, poor drainage around the foundation, or deterioration of the masonry. All these conditions call for remedy by a professional before you attempt to correct any problems they may have caused in the floors themselves.

Leveling Sagging Floors   Once you have determined that none of the foregoing problems exist, there are various steps you can take to correct certain sags and slopes. Joists may be reinforced and girders may be raised, but first you must locate the exact problem area. Find the low spot by allowing marbles to roll or by laying a long, known straightedge, such as an 8-foot length of rigid pipe or the factory edge of a strip of plywood, on the floor.

When you locate the low spot from above, you'll have to relocate it from below. Note the spot's relation to walls, heating ducts, or pipes that may pass through the floor—anything that will help you identify it when you get underneath.

You may find that the low spot is located over a joist that has sagged, or in the area of a girder or its supporting post. If the joists and girders appear sound, check any girder supports, especially where they rest on the slab. Cracks in the slab, if not too large, are usually not serious and can be simply patched (see page 68), but any substantial tilting of the slab or large cracks are signs that the slab was poured on improperly prepared ground. This should be checked out by an engineer.

If the low point is in the area of a joist, check the joist's condition. If it is discolored and near plumbing, it may have suffered water damage and begun to rot. It will need to be replaced. If there are signs of insect infestation—suspicious holes or little piles of dust, or if the wood feels spongy to a sharp probe such as a penknife—then it may also need to be replaced. If neither of these conditions is apparent, the joist may just have sagged through weakness or warping. You can attempt to raise it (see below). Check the neighboring joists and the nearest girder, including the girder supports—posts, piers, and the points at which the girder rests on the foundation wall. If you are at all unsure about the cause of the problem, consult a reputable professional.

Jacking Up the House   Once assured that the house is structurally sound and that any serious defects have been professionally remedied, there are several ways you can try to level the floor yourself, providing the sag is not more than an inch or so. You can either strengthen the existing joists or add supports. In either case you will first need to raise the offending area. Jacking up a house and modifying the joist structure is hard labor, fairly difficult work, and a serious undertaking. It can take much more time and effort than anticipated. An inexperienced person can run into difficulties and make an error in judgment. If you're uncommitted to the task or uncertain of your skills, have a professional do the work.

How you jack up the house will depend on how much space there is below the floor. If you are in a full-height area, such as a basement, you'll be able to use adjustable metal *jack posts* or wooden posts on a *house jack* (see Figure 3). If there is less head room, such as in a crawl space, your only choice is a house jack. Whether you use a jack post or a house jack, it should be placed on a stout wood pad or cribbing to distribute the weight it will carry. Greasing the threads of either type of jack will make it far easier to use. Before you install the jack or jack post, check to see how many turns of the handle cause it to rise $1/16$ inch. That's the amount you'll raise the jack at any one time. Note that adjustment of the jack by screwing is what differentiates a house jack from a regular hydraulic jack. With a lever-operated hydraulic jack it would be too hard to achieve the small, carefully controlled adjustments needed to jack up a house.

Positioning a Jack Post or House Jack   A metal *jack post* is usually fitted with top and bottom plates that can be nailed to the wood pads at its bottom and to the heavier timbers at its top that are placed beneath the area to be raised. *Duplex nails,* nails with double heads, should be used so that they can be easily removed. Once you've provided a secure base for the jack post, screw the adjustable top of the post all the way in to allow yourself

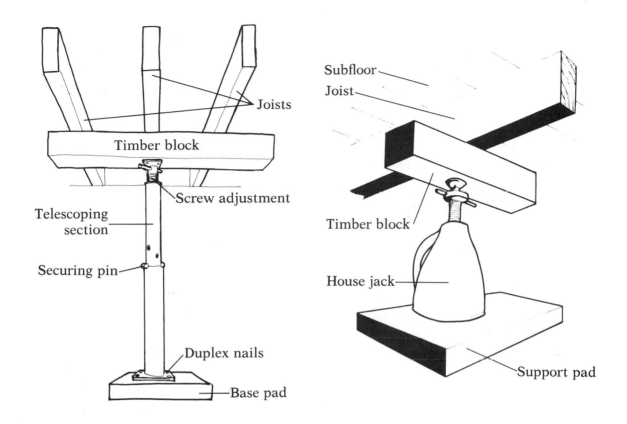

Using an adjustable metal jack post   **Using a house jack in a crawl space**

Figure 3   Jack post and house jack

the maximum available extension later. Now extend the post—the top section rises out of the bottom section and is held in place by a steel pin fed through holes in opposite sides. Extend the post so that it reaches the bottom of the timber you are using under the joists or girder in question. Before fixing it finally in position, use a level to ensure that the post is perfectly vertical. If it is not, it could slip out sideways when you put it under pressure. You position a house jack the same way, only it does not require nails to secure it.

If you are attempting to raise one joist only—because its fellows do not need raising—it is still a good idea to place a stout timber between the top of the jack and the bottom of the joist in order to spread the effect over a wide area. If you were to concentrate it on one spot, you risk crushing the joist at that spot rather than raising it. It is far more likely, however, that several joists will need raising. One stout timber placed in such a way that it rests under all of them is the most efficient way to raise all the joists at once.

Raising the Jack   Once the jack post, or jack, is in position, you can begin to raise it. Do not be impatient; raise it no more than $1/16$ inch in a 24-hour period. That will allow the structure of the house to adjust to the new pressures being put on it without causing plaster to crack or trim to burst loose or the dislocation of a plumbing line.

Keep an eye on the base of the jack to see that the pressure exerted is not causing the jack to be pushed into the ground below rather than raising the floor above. If it is, you must start again, this time providing a broader base.

Reinforcing a Joist   If you're raising the floor in order to reinforce a joist that appears too weak or has warped or shrunk, you'll need to use the jack under a timber large enough to span several joists on either side of the joist to be reinforced. Place small blocks of two-by-sixes between the bottom of each joist and the timber so there'll be room to slide in a new length of joist between the floor and the supporting timber. And before jacking up the timber, remove the bridging on either side of the joist to be reinforced so the new length of joist can be butted directly against the weak one.

Raise the flooring using a jack, as just described, until you are satisfied with the level achieved, then go a little higher to allow for slight settling. The easiest way to reinforce a joist is to secure another length of the same size material to the side of the newly leveled joist. The new piece should be somewhat longer than the sag was but doesn't have to equal the original span. Whatever the size called for, choose the straightest specimen you can find and sight along both edges to determine which, if any, has a crown to it—that is, the convex side—and install it uppermost.

If dampness or insect damage played any part in weakening the old joist, treat the new piece—and the surface of the joist to be reinforced—with wood preservative. Resist the temptation to cut out sections of the new piece to accommodate any pipes, wires, or ducting that may interfere. That will only weaken the support, and may, in fact, be the reason why the original joist has sagged. Instead, reroute the obstructions, getting professional help if necessary.

Attaching the New Piece   Securing the new piece to the old will require a helper, and even so may be difficult. If there is room to swing a hammer, nail the piece to the original joist using 16-penny nails, staggered top and bottom about 12 inches apart. Should they protrude, clinch them by bending over the ends that stick out. It may be easier, though a little more time consuming, to hold the joist in place with large C-clamps and drill holes for bolts. Use at least ½-inch bolts with washers. Bolts may be spaced a little further apart.

Reinstall any bridging you may have removed, reducing its size if necessary, and then lower the jack in the same manner as you raised it—bit by bit. If the floor sinks below the desired level you must either install a longer reinforcement or reinforce the adjacent joists as well.

The procedure described assumes you are dealing with a sag on the ground floor. If an upper floor sags, the cure may be more difficult. If the sag is near a partition wall, and this partition wall is directly over a similar partition wall downstairs, then leveling the lower floor (possibly by installing a new post, as described in the next section) may well take care of the upstairs problem as well. If the sag is in the center of an unsupported area, the cure is more complicated, since the sagging floor has a finished ceiling beneath it. The same method of repair will work but part (at least) of the finished ceiling must be removed before you can begin. Before considering this, read the section on ceiling construction that starts on page 177.

Installing a Permanent Post   If the sag involves more than just one or two joists, the solution may be to raise the height of the girder, thus raising *all* the joists. This may be necessary because there are insufficient supports or because those already there may have sunk, due to deterioration.

The additional or replacement support can be a wood post of the same dimensions as the existing posts, a prefabricated metal post, or, if local building codes permit, an adjustable metal jack post such as we described on page 14. As with jacking a house and reinforcing a joist, installing a post is a serious and fairly difficult undertaking. Use a professional if you have any doubts about the job.

If you are replacing an existing post, do not remove it without supporting the girder temporarily on either side. Raise the temporary supports slowly until you are sure all the weight has been removed from the original support. Once again, be patient and do not raise the girder more than $1/16$ inch each day.

Before installing either a replacement post or a new one, you must provide a secure base. Deterioration of the previous post's footing could have caused the sagging floor in the first place, particularly if the post still seems to be in good condition. A post that rests on a slab should have its own footing in the slab. If the footing has deteriorated, consult an engineer. He or she can determine the correct size of the new footing, and a construction professional should install it.

With the girder raised to its proper level, measure and prepare carefully the length of post required. If you are installing a metal jack post, you will be able to extend the jack to fit. If you are installing a wood post, it will help to raise the girder a little higher so that a post of the proper height can be slipped into place. Secure the bottom of the support first, nailing it into place on its pad. Then use a level to make sure that it is perfectly vertical and lower the support jacks so that the new post takes the weight—but once again, do this in small daily increments. Finally, secure the top of the new support using a *metal post cap* for a wooden post or bolting through the flange provided on metal posts.

Tightening Springy Floors

Floors that give when you walk on them can be firmed up by building up the support underneath. One common cause of springy floors is lack of sufficient bridging between the joists. Adding a row of solid bridging in a line across the middle of unsupported or unbridged spans is an easy fix, and the job requires no special tools.

Solid bridging should be made from material of the same dimension as the joists themselves, cut to fit snugly between adjacent joists and toenailed into place. It is best to fix the bridging in one aligned row, although in some houses it is often staggered to make nailing through the joists into the ends of the bridging pieces easier. Bridging is somewhat difficult to install as a retrofit since the flooring on top of the joists makes nailing the top ends of the bridging awkward. Metal-strap bridging is not so awkward to install since the ends are flattened into small flanges, to simplify the nailing.

A springy floor in an older house is usually the result of too few joists or joists spaced too far apart. To tighten up such a floor, install additional joists or double up existing joists by nailing new ones to the sides of the old ones. An alternative cure is to install an extra girder that will support all the joists at the middle of the span. You don't have to support the new girder by building it into the foundation walls—a tough job. Simply support it at either end and at the center with wood posts or metal jack posts (see previous section).

Eliminating Squeaks

A squeak is invariably produced by adjacent pieces of wood not held tightly to each other. It can be traced either to the finish floor itself or to the underlying structure. There are many possible causes, but the cures are relatively simple and can be accomplished by any reasonably handy person. When you consider the number of parts that make up a flooring system, and all the things that can go wrong, a squeak that is simply an annoyance to you can reveal a great deal about the state of your floors.

If you have other problems with your floors—troublesome joists, bridging, girders, and supports—the repairs you make may automatically eliminate many noises and squeaks.

Any remaining squeaks must be tracked down one by one. If the underside of the squeaky floor is accessible—such as in the basement—have a friend walk across the floor as you watch from underneath. When the floor squeaks, you should be able to locate the spot and, usually, the cause of the squeak. If the cause is springy joists or loose bridging, the cures have already been covered. More likely, the squeak results from flooring no longer securely fastened to the joists.

Using Shims and Cleats   One way to eliminate squeaks caused by loose flooring is to tap *shims* (made from shimming shingles, bought in small bundles at lumberyards) between the loose flooring and its supporting joist (see Figure 4). This works best when the subflooring consists of diagonally laid boards, since these boards frequently dry and shrink unevenly, resulting in squeak-producing gaps. You can shim one or two boards, but extensive or too-vigorous shimming may simply push the subflooring away from the joists.

If a whole section of flooring appears to be loose, a better method is to nail a *cleat*—a length of one-by-three or bigger—to the joist immediately under the loose section. To get the cleat really tight against the flooring above before nailing, cut a post (a two-by-four is perfect) slightly longer than the distance between the cleat and the floor. Hammer the post snugly under the cleat. Again, be sure not to force the flooring even further away from the

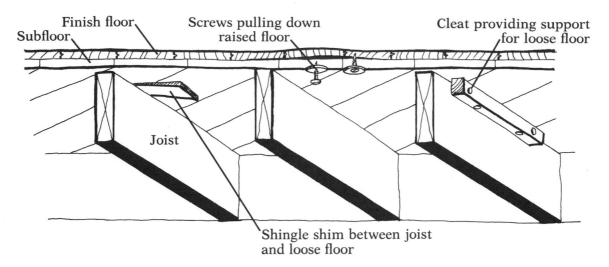

Finish floor,

Subfloor,

Screws pulling down raised floor,

Cleat providing support for loose floor

Joist

Shingle shim between joist and loose floor

Figure 4   Cures for loose and squeaky floors

joist; you simply want to close the gap. Nail the cleat to the joist. Then remove the post, and nail up into the subfloor (Figure 4).

When plywood or other types of composition board are used as subflooring, the sheets are usually arranged with their long side at right angles to the joists. This can leave unsupported joints in the subflooring between joists, and occasionally at other spots. (High-quality subflooring made with interlocking edges eliminates this problem.) If you spot movement at these unsupported joints when someone walks on the floor above, there are two possible fixes: Either insert an additional piece of solid bridging under the joint or attach a piece of one-by-four beneath the joint, screwing it to the sheets on both sides of the joint.

Tightening the Floor with Screws   If the squeak is caused by loose finish flooring rubbing against subflooring, you may not be able to see anything from below. But if you can pinpoint the trouble spot, you may be able to cure it by pulling the finish floor down tightly against the subflooring with screws (Figure 4). This works better in theory than it does in practice, but it's worth a try. (If the finish floor has lifted away from the subfloor for some obvious reason, this should be corrected first—see the next section on repairing buckled flooring.)

Start by flattening the loose finish floor as much as possible by having someone stand on it. Then drill pilot holes for the screws. It is best to use the fattest screws possible but be sure that they are at least ¼ inch shorter than the combined thickness of subfloor and finish floor, which is usually about 1½ inches. Use roundhead screws with a washer under the head, which will prevent the screw from crushing the wood and being drawn up into it.

The Squeaks Upstairs   If the squeak is in an upper floor, with an underside that is finished or otherwise inaccessible, you have to work on the floor's surface. You can sometimes wedge loose boards together by driving glaziers' points between them but the points may work themselves out again, a hazard to bare feet.

A safer method is to try lubrication. *Reoiling* an oil-finished floor may help, as the oil will sink into the floor and help expand the wood. Floor oil between the cracks of any wood floor may also work, as may powdered graphite or even talcum powder. These solutions can be messy, and certain oils may stain the floor. Try some first in a closet or other inconspicuous area.

If these simple methods fail and you're sure the problem is in the floorboards, you can try to toenail or screw the boards down to the subfloor (Figure 5). Use ringed flooring or annular nails in pairs, so they slant toward each other. Drill smaller pilot holes first, since hardwood flooring is generally too dense to nail. Set the nailheads below the surface and fill the holes with wood filler.

The more sophisticated way is to use screws. This may well have been how the floor was

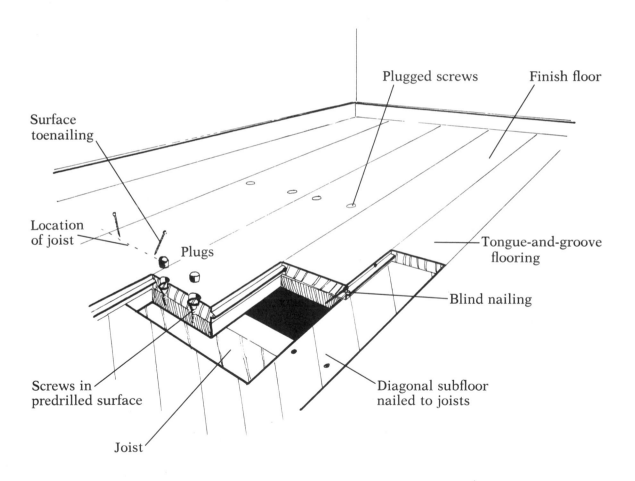

Figure 5   Repair of finish floor

originally secured, so any repairs will be less noticeable. Once again, drill proper pilot holes for the screws—using a bit the same size as the screw's shank to start, then a smaller bit for the threaded section of the screw (combination bits for various types of screws are also available). Take care not to drill too deeply. Apply some weight to the boards when you insert the screws, which should end up ¼ inch below the surface. Finally, plug the screw holes with wood *plugs,* available at lumberyards or hardware stores (Figure 5). You can also make your own plugs with a simple plug-cutter bit in your drill—use a piece of scrap flooring similar to the floor you're repairing. Cut the plugs so they protrude slightly from the hole and sand them down.

Replacing a Floor Section

A floor's surface sometimes looks more seriously damaged than it really is; it often can be repaired by refinishing alone. However, if it is really badly damaged—with deep gouges, deep burns, cracked or split boards, or swollen and buckled boards—you'll have to remove and replace the damaged section. In many cases, however, matching the old material exactly may be very difficult, if not impossible.

Hardwood floors can be divided into three main types: *wide board flooring* (of soft or hard woods), *narrow strip flooring* (most commonly of various types of oak, maple, or pecan), and *block* or *parquet flooring* (made of a large variety of woods).

Wide-board flooring is the oldest type and the easiest to repair—it is unlikely that you'll have to remove more than one board at a time. Very old board flooring was sometimes made with square edges and attached simply by being nailed through the surface. More often, however, these boards had edges that fit together, and were nailed at the joint so that the nail was concealed.

Strip flooring is composed of narrow boards, 2 to 4 inches wide, that are invariably made with a tongue-and-groove joint not only along the long sides but at the ends as well. This allows use of odd lengths without having to position the end of a board over a joist, as is the case with wide-board flooring. You can therefore replace any length of strip flooring you like, but when replacing wide boards, the replacement piece must be cut so that both ends finish over a joist. Replacing flooring requires some skill and much diligence. To achieve good results, proceed with care. In addition to some basic tools, you'll need either a drill with large bits and a chisel or a power circular saw.

Removing Old Flooring   Whether you remove one wide board or several narrow strips, the procedure is the same. Cut the board or boards squarely across at each end of the section to be removed, then split out the damaged piece or pieces, starting at the center. Here's how to go about it: To make the square cuts at the ends, you can use a drill and then a chisel; drill a hole near the corner of the board and cut the rest away with the chisel. It will be a far better job, however, if you use a circular saw. Tack a length of lath or batten to the floor as a guide, using a try-square (see Glossary) to fix this guide at a perfect right angle to the length of the floorboard. Set the depth of cut on the saw to the exact thickness of the floorboard (establish this by measuring a cross-section, in a closet for example, or under a register or grill cover plate, or even remove a piece of baseboard, if necessary).

Do not saw all the way across the width of the board; stop ½ inch from the edge. There are two reasons for this: You will not risk sawing through any nails that hold the edges of the board down; you will not cut off any projecting tongue or groove from an adjacent board. Cut through these corners with a chisel, split them, and pry out the damaged section and carefully remove the remaining edges.

When you are dealing with a damaged section of strip flooring, it may seem easier to remove a complete rectangle, but if several boards are involved, a staggered replacement will result in a better matched repair.

Installing New Material   When the damaged area is clean and there are no nails remaining around the edges, cut the replacement material to length very exactly. Insert one board after another so that the tongued edge of each can be blind-nailed, as shown in Figure 5. When you come to the last board (or if only one board is to be replaced), the procedure is a little different. Either you can cut off the projecting tongue of the existing board or you can cut off the bottom part of the groove on the replacement board and, after inserting it, face-nail the final board down. Either of these methods will allow you to cope with replacement boards that are not exactly the same width as the boards they are replacing. However, it will look better if you are able to keep the strips aligned.

When the damaged area has been filled, set any visible nails below the surface of the floorboards and fill the holes with wood filler. Before refinishing the area to match the surrounding floor, you may have to sand the replacement boards so that they are level with the rest of the floor. Refinishing is discussed in the next section.

The replacement of parquet or block flooring may be easier since the blocks, though usually tongue and groove, are often merely glued into place. If you can find the right solvent for the glue used (from the original supplier or contractor, or perhaps from a flooring supplier), removal will be easy. When using solvents, work with the room's doors and windows open and create cross ventilation with a window fan set to exhaust air. It won't be easy, however, to find exact replacement blocks. Even if you do, chances are that the color won't match the existing floor—time and previous refinishings will have altered its color. Settle for the closest you can find and try to match the replacement by experimental staining on an extra block.

If you have to saw out pieces of block flooring, follow the procedure described above for strip flooring to remove the pieces without damaging the edges of the surrounding blocks. To replace a similarly sized block, simply cut off the bottom parts of the grooves on the replacement block, and glue the block into place. Tap it firmly into place with a hammer (protecting the flooring with a piece of scrap wood), and clean up any excess glue quickly before it mars the surrounding wood.

For repairs that involve the removal and replacement of baseboards or door trim, see chapter 16, Trim.

REFINISHING A WOOD FLOOR

When the traditional wood floor finishes, such as varnishes, shellacs, and lacquers, become worn and dulled to the point where no amount of polishing can bring back a shine, they must be completely refinished. Most floor finishes can be redone simply by buffing with steel wool, wiping with mineral spirits, and applying another coat.

If the floor has not dulled, there still may be good reasons for resurfacing: to restore heavily damaged floors, or to match floors that have been partially carpeted, have worn unevenly, or have had unknown finishes applied to them. Although not a difficult job, resurfacing a floor involves a lot of work, and takes a considerable amount of time and some skill and care. Be sure you understand what has to be done before you commit yourself to doing the job.

Basically, resurfacing involves three major operations: preparation, sanding, and application. Preparation is relatively simple, but when sanding, you will be using a machine that is a bit tricky to operate. You can do an excellent job if you're careful. Application involves selecting and applying the new finish, again with careful attention to detail.

Note that there is little point in refinishing a floor that has structural problems or serious surface damage. No matter how well you shine the floor surface, you may simply have to repeat the process if you have to repair the floor in any way. Before starting, therefore, make a thorough inspection of the floor's condition, rectify any structural problems, and repair any surface blemishes, as outlined in previous sections of this chapter.

Preparing the Floor for Sanding

Prepare carefully and well before running any sanding machine. You must empty the room, take down curtains or drapes, and remove any floor grilles, registers, or gratings. If there are radiators in the room, remove the collarlike flanges that are fitted around the pipes where they pass through the floor. If there are any floor openings, such as holes left under removed grilles, make sure these are temporarily but securely blocked off. Although most floor sanders have dust bags, there will inevitably be a lot of dust created, dust so fine that it will find its way into any unprotected opening or crevice. Doorways and all other interior openings should be covered in plastic with the edges of the plastic sealed tightly with masking tape. But open all windows and observe all other precautions described in Personal Safety.

One of the hardest parts of sanding a floor is getting into corners and up to edges. You can make this a little easier if you remove the quarter-round strip, called *shoe molding*, which may cover the joint where the floor meets the baseboard. Do this carefully, using a broad chisel to pry the molding away just a little. Insert wedges to hold it away from the baseboard while you pry the next section. If you do this carefully enough, you will be able to save the molding and reapply it after the floor is finished. Make sure you remove any nails that may be left protruding. Check the floor itself carefully—board by board—for any other nails or staples projecting above the floor surface. Pull them out if you can. If you can't, use a nailset to sink them at least ⅛ inch below the surface, filling the holes with wood filler (match the color of the existing wood, if you can).

You are now ready to sand off the old finish.

Getting Properly Equipped

Unless you intend making a career out of refinishing, you should rent, not buy, a floor sander. It's a fairly big machine, expensive, noisy, and not the sort of thing you want to store just to refinish another floor at a later date. Make sure the renting agency explains the sander's operation before you take it home. Sanders are not complicated, but there are certain adjustments you will have to make, including changing the sandpaper. Make sure you come away with the necessary instruction booklet and auxiliary tools, such as wrenches and keys. A test run at the rental agency is very helpful, if it can be done.

One of the obvious signs of a poorly refinished floor are wavelike gouges caused by inexpert sanding. These can be avoided to a large extent if the sander you rent is designed so that you can raise the drum from the floor by using a tilt-up lever rather than by having you lift the machine itself.

Along with the big drum sander, you will also need an edge sander—a much smaller machine—used, as its name implies, for sanding around the edges where the big drum sander cannot easily reach.

The Right Sandpaper For both machines you will need an adequate supply of sandpaper of different grits; the lower the grit number, the coarser the paper. Since most rental shops will only charge you for the paper you actually use, take more than you think you'll need. How much, and which grit, depends on the size and condition of the floor: for really bad floors, perhaps painted and gouged, you should start with 20-grit paper; for varnished floors in fair condition, 30-grit to 40-grit paper is adequate.

After the first sanding you will repeat the process with a medium grit—possibly a 60-grit paper—and finally finish with a third sanding using an 80-grit or 100-grit paper. As a rule of thumb, for every 200 square feet of floor to be sanded you can estimate at least two belts of each grit for use in the floor sander, and up to four sheets of each grit for the edge sander. But allow for unexpected disasters and provide yourself with more, just in case.

Personal Safety For safety reasons—because of the noise and dust created—use ear plugs and a dust mask. Wear old clothing, of course, and pick a pair of shoes that will not leave marks on the newly sanded floor (the raw wood surface is very vulnerable until you get it refinished). Exposed flame and sparks can cause fine sawdust suspended in the air to explode.

To prevent fine particles from building up, ventilate the room well, opening all windows and using a window fan set to exhaust air. As an extra safety precaution, don't smoke. Continue running the window fan for a half hour or so after you finish sanding. One added safety precaution: Make all adjustments to the sander, including changing the paper, only when the machine is unplugged.

When you're sure you're properly equipped and the floor is clear of all obstructions, you are ready to start sanding. But start the job only if you'll have time to put on the first coat of finish that day. The raw wood is extremely vulnerable to moisture and might swell or at the least suffer from raised grain if left too long before being finished. Either problem will mar the finish.

Working with the Sander

Raise the drum from the floor, start the machine, then lower the drum and start to move *instantly,* always going with the grain—that is, along the length of strip or board floors. If you are working on block or parquet floors, where there is no single grain, compromise by sanding diagonally. With badly damaged floors and cupped boards it may be easier to work diagonally across the boards at first, working away the high spots until an overall level is attained (then work with the grain).

When you stop the sander, lift the drum from the floor *before* you come to a halt. It takes only a second of standing in one spot with the machine running for it to create an unsightly groove. Having sanded one strip, return on the next strip with a few inches overlap. The idea is to remove only the finish and as little wood as possible: keep moving and work as evenly as you can. When the first sanding (called the first *cut*) with the drum sander is complete, sand any remaining areas with the edge sander, using the same grit paper.

For areas that even the edge sander may not be able to reach, such as behind radiators or pipes, use a hand paint scraper to remove as much of the finish as possible. If this doesn't work, wrap a small piece of sandpaper over a small wood block and work by hand. As soon as this is all done, it is time to switch to the next finer grit paper for the second cut. Before starting, check to see whether the first cut has exposed any nails. If so, set them and fill the holes. If you are sanding diagonally on a block or parquet floor, work each successive cut on the opposite diagonal.

After the second cut is complete, check again for nails, and to see that all areas have been sanded equally and no grooves have been created. The final cut, with the finest paper, should be taken very carefully. Any gouges you make at this point will have to be removed by a heavier grit. Make a final, close visual inspection, looking most carefully at the corners and around obstructions such as pipes, radiators, and vents. Work on these areas by hand if they need touching up.

Cleaning Up Before even thinking about applying the finish, give the dust time to settle, then clean it up with a vacuum cleaner and a damp cloth. Ventilate the room well; turn on the room air conditioner, if there is one, or use a window fan set to exhaust air. Check again for any blemishes, such as unfilled nail holes or scuff marks. Any marks you miss will be preserved for a long time under the new finish.

A word of caution concerning dust disposal: Since the sanding machines generate a certain amount of heat, and since what will have been removed may contain residues of varnish, lacquer, or other inflammable materials, the dust can self-ignite. Allow the dust to cool before containing it in any closed space.

Selecting and Applying a Finish

You have essentially two choices regarding what kind of finish to use. Apart from coloring the wood with one of a variety of stains, you can apply a penetrating finish or a surface finish. Penetrating finishes actually sink into the wood, sealing it. They wear only as the wood wears, enhance the richness of the wood grain, and are easily restored, should odd spots require touching up. They usually go on in two coats. Penetrating finishes may require subsequent waxing or cleaning.

Surface finishes form a skin or film over the wood, and although often extremely durable (as is the case with polyurethane varnishes), the film is eventually subject to wear.

Not all finishes can be used in conjunction with all stains or surface finishes—read labels carefully. Note that flooring laid with adhesive—such as block flooring—should be finished only with materials that are compatible. Chemical reaction and degradation of the adhesive are possible.

There are so many products available that it is difficult to generalize on application procedures, but a few points may be noted. Before starting to apply any finish, plan the sequence of application: Do not "paint yourself into a corner"! Work in an adequately ventilated room; establish cross ventilation with open doors and windows and a window fan set to exhaust. Wear a long-sleeved shirt and rubber gloves. Before starting, turn off the range, heating system pilot lights, and other open flames, even those in neighboring rooms or down a flight of stairs. Don't smoke, and don't use electrical equipment or anything else that could produce a spark or a flame.

If you are using a penetrating sealer, it must usually be applied liberally at first and allowed to penetrate. Then the excess is wiped off. If you are applying a surface finish, read the directions concerning drying times, especially as they vary with temperature and weather conditions. Be prepared to rub down intermediate coats, most easily done with a rented floor buffer that uses a pad of steel wool. Like the floor sander, the floor buffer may have difficulty reaching certain areas, so these must be finished by hand. After you have used a buffer, the floor must be immaculately cleansed of all dust, grit, and residue from the steel wool before a subsequent coat of finish is applied.

Polyurethane varnishes, which have largely superseded other types of surface finishes, may be waxed, though many manufacturers maintain this is unnecessary and precludes the opportunity to apply a subsequent coat later on. According to the manufacturers, any wax on the surface will cause added polyurethane to "fish-eye" and not form an even coat. Wax can be removed by buffing with steel wool, then wiping the floor with mineral spirits.

One of the most difficult parts of the floor-refinishing process is having the patience to wait until the finish is completely dry before replacing grilles, moldings, carpets, and furniture. Adhere strictly to the manufacturer's instructions on drying time. After the work you've done, the job should last a long time. Don't mar it at the outset by being overly eager to put the room back in shape.

3

RESILIENT FLOORS

ASSESSING THE CONDITION

Under the heading of resilient flooring is a wide range of materials, from cork, asphalt, rubber, and the no-longer-common linoleum to the currently popular vinyl and vinyl composition. Whether these materials come in sheets or in tiles, they are all among the easiest of floor coverings to install or repair. They can create an almost instantaneous new look, especially since they are now available in an array of patterns and colors.

Repairs to an existing resilient flooring, whether laid in sheets or in tiles, are relatively easy. Many of the repairs, however, are necessitated by conditions beyond the flooring itself. You must therefore determine the cause before proceeding with the repair.

The condition of the underlying floor structure can seriously affect the appearance of resilient flooring, so any obvious faults—unevenness, sloping, and squeaking—should be addressed first, as described in chapter 2.

Underlayment Problems

Small bumps all over the floor may be the protruding heads of the nails in the floor layer beneath. Resilient flooring should be laid on as smooth and as level a surface as possible. This is usually achieved by providing an underlayment, in the form of sheets of plywood or composition board. In new construction the underlayment is often applied directly to the floor joists; in older construction or in renovations, the underlayment is often installed over an existing floor.

Nails holding down the underlayment may pop up for several reasons. The wrong type of nails may have been used. *Cement-coated nails* or *annular nails* with ringed shanks provide more gripping power than other nails and should have been used. Even well-constructed floors have a certain amount of spring or give, and this slight but constant flexing can cause regular nails to work their way out. Of course, the more flexible the floor, the greater the likelihood of this happening. A floor with too much spring may well have weakened or insufficient joists or girders; this must be fixed before dealing with protruding nails, or the problem will only continue.

Underlayment that has shrunk will also result in protruding nail heads, as well as in a related problem: indentations visible in the surface of the floor. The problem may be the result of individual boards in the subfloor drying out, or of changes in the subfloor structure due to settling of the house. In either case, the underlayment, whatever it is made of, is pulled out of shape just enough to force the nails up or produce gaps under the flooring. These show up as slight grooves in the finish flooring, over the joints of either the underlayment sheets or the boards of the subfloor.

The opposite condition can also cause problems. If the underlayment swells or buckles, for whatever reason (most commonly because of moisture), the resilient floor covering is also going to buckle. Since the floor covering may buckle for reasons of its own, you must first check the condition of the underlying structure before dealing with the floor covering itself.

Moisture, which can result from poor drainage or plumbing problems, is the major reason adhesive fails. Before trying to restick loose or curled sections of flooring, make sure that there is no leaking nearby and that the floor underneath is free from moisture, whatever the cause.

Unfortunately, all of these problems call for the floor covering to be removed and the underlying problem to be fixed first. Merely repairing or replacing the resilient flooring itself will not eliminate the condition.

REPAIRING AN EXISTING FLOOR

Since complete replacement of a resilient floor is relatively easy, it is not worth making extensive repairs in an existing floor. Repairs should be limited to the flattening of bumps or blisters (those not caused by faults in the subfloor), regluing raised edges, replacing individual damaged tiles, and replacing sections of sheet flooring. By exercising reasonable care, anyone, even someone who has limited skills and experience, can make these repairs and achieve good results. Nevertheless, whether you are repairing or laying resilient tile, when you work with adhesives or solvents, open the room's doors and windows and ventilate the area with a window fan set to exhaust air.

Availability of Matching Material

When you're assessing the condition of your flooring and a possible fix, consider the availability of matching material. If the flooring has been recently installed, you may well have some leftover material, or your supplier may be able to provide some replacement pieces. Since there are so many shapes, colors, and patterns of resilient flooring, a perfect match can often be a problem. One solution is to consider a contrasting color or pattern for a specific area—this often looks better than a failed attempt at a perfect match. Unfortunately, if you've been hoarding extra material for a long time, it may no longer match what has been installed because colors fade with time.

Bumps and Blisters

To flatten a small bump or blister, cut it across the middle with a sharp knife (a sturdy utility knife or a linoleum knife), smear adhesive under the cut, and apply a heavy weight to the area until the adhesive dries. Remove any adhesive that oozes out of the cut as the weight is applied. Which adhesive to use depends on the type of material on the floor. If you know what it is, you can get the answer from a dealer or supplier of that flooring, or by lots of label reading. If you don't know, take a piece of it to the dealer. Once you decide on an adhesive, the questions of which solvent will clean up the excess and how long it will take to dry will be answered by the label directions.

Regluing Loose Flooring

Before trying to reglue peeling, curling, or lifting flooring, make sure that the cause of the problem has been cured. It's also important to remove as much of the old adhesive as possible. Some adhesives will work on top of old, but many will not. It's better to be sure, and start with as clean a surface as possible.

Replacing Individual Tiles

When you want to replace a damaged or broken tile, the main problem is to remove the old tile cleanly. If it's already started to lift, help it out with a putty knife and a little heat. Vinyl and rubber tiles can be loosened with the help of a hot iron—use a towel between the iron and the tile. Heat softens the tile and the glue holding it. If that doesn't work, use a combination of cutting, chipping, and scraping. The important thing is not to damage the adjacent tiles and, again, to remove as much of the old adhesive as possible. Do *not* sand tiles that contain asbestos, since this can create airborne asbestos particles that can harm your health.

When the area is clean, make sure there are no protruding nails and that the area is level; fill any holes or cracks with wood filler. Test the replacement tile for fit—is it exactly the right size and shape? Does the grain (even random-pattern tiles often have a grain) go the right way? Now apply the proper adhesive, as noted before, to the new tile. (Also have the right solvent on hand.)

Insert the new tile by aligning one edge with the matching edge in the vacant spot. Press the new tile against the old as you lower it, using the aligned edge like a hinge, into the space. Press it down from the center outward and clean up any adhesive that seeps out from the edges. If it is level, apply some substantial weight—20 pounds of books or bricks is usually sufficient—and allow the adhesive to dry for the period called for on the adhesive container.

Replacing a Section of Sheet Flooring

Replacing a small area of sheet flooring is not much harder than replacing an individual tile. The biggest problem will be to match the pattern of the replacement piece to the existing floor. You can do this perfectly, using the professional's technique of *double-cutting*. Here's how it's done (see Figure 6).

Start by cutting out a replacement piece larger than the damaged area and position it over that area so that the pattern of the new piece aligns with the pattern on the floor. Now tape the replacement piece securely in place so it won't slip under pressure. Mark out

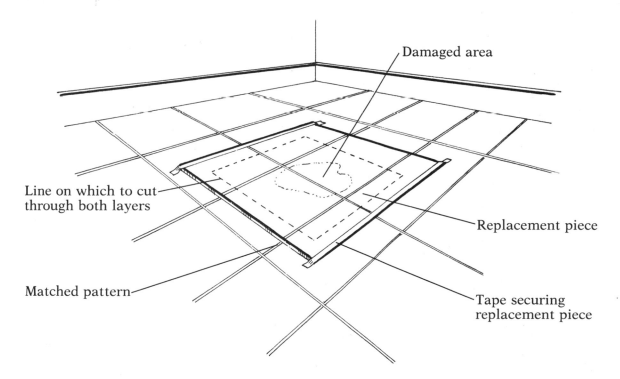

Figure 6   Patching sheet flooring

a patch large enough to cover the damaged area. The patch will usually look better if it is rectangular or even square, and if its edges align logically with some part of the pattern in the flooring. To be sure of a square cut, use a carpenter's framing square as your straightedge, since this tool has two arms at right angles to one another.

Now, with a utility knife (use a new blade) and a metal straightedge, cut through *both* the replacement piece and the underlying material at the same time. This takes a strong hand and will doubtless require repeated cuts before you cut all the way through.

The patch is now cut and the damaged area outlined by a cut line. The procedure is now the same as for replacing an individual tile. Thoroughly clean the damaged area, chipping or scraping out all traces of the original material and cleaning away as much of the old adhesive as possible. As with tiles, do *not* sand sheet flooring if it contains asbestos. The airborne asbestos particles that sanding can create can harm your health. Apply new adhesive (with the same cautions about type and solvent as before). Insert the replacement piece, applying pressure from the center outward, and cover the area with something heavy. Wait for the adhesive to cure.

INSTALLING NEW RESILIENT FLOORING

Selecting the Material

In choosing new material for a floor, you have to decide between *sheet flooring* and *tiles*. Most materials and styles are available in either form. Installing a new resilient floor does not require a great deal of skill, but it does demand a fair amount of time and hard labor. In particular, surface preparation requires effort and more time than many people anticipate. Sheet flooring is more difficult to handle because of its size—it usually comes in rolls 6 to 12 feet wide—but it is quicker to install than tiles, which require rather more careful layout. For relatively small rooms, such as kitchens and bathrooms, an almost seamless floor is a great advantage.

Once you've decided on sheet or tile, you'll have to choose among materials, some of which may be better suited to one application than another. Visit several suppliers to get a better idea of the choices available.

The heaviest vinyl floorings are the inlaid vinyls, and because of their weight, cost, and increased unwieldiness in sheet form, it is probably better to let professionals install this type. Lighter, thinner, and more flexible types are much easier to handle and can be installed by the average do-it-yourselfer.

Tile flooring most be glued in place. But even here you have a choice: Some tile flooring is manufactured with a self-adhesive surface (covered by a peel-away backing), while other varieties require the application of a separate adhesive. Resilient tiles are most commonly found in 12-inch squares, but it is possible to find other sizes.

Whatever floor material you choose, a floor plan—an accurately dimensioned drawing of the floor—will help you estimate how much you'll need. Your supplier will advise you how best to arrange adjacent sheets if this is what you're using, and may even be able to precut some sections for you. If you're using tiles, you can arrange your own pattern, which is most easily done on scaled graph paper. In either case, when you are purchasing allow at least 15 percent extra for waste, plus some for possible repair use later. The larger the area to be covered, the smaller the percentage need be. Often it's possible to return unused material. If the dealer allows returns, allow a full 20 percent for waste, but check with the dealer first.

Let the new flooring acclimate itself to your house for a day or so after it is delivered. Some materials expand or contract, depending on temperature. Storing it in the room where it is to be used is best.

Preparation of the Subfloor

Since most forms of resilient flooring are relatively pliable, they will conform well to the surface of the floor. That also means they will show any irregularities in that surface— gaps, holes, ridges, protruding nails, or other blemishes. It is extremely important, therefore, to lay resilient floors on a base that is perfectly smooth. Some form of underlayment is required.

Preparing Masonry Floors   You can successfully install resilient flooring over concrete if the slab—such as a basement floor—is perfectly dry and completely free of oil or grease. Any moisture may damage the flooring or cause the adhesive to fail. Freshly poured concrete takes a long time to cure completely—often a month or more. Wait until the concrete has cured, even if it appears dry.

If there are any moisture problems, the concrete should be treated with a waterproof sealant first. Oil or grease deposits should be cleaned off using commercial preparations available at hardware stores (check the precautions carefully). Once cleaned and dry, the surface must be smoothed. Any cracks or irregularities should be filled with a latex filler designed for the job, also available at hardware stores (see page 69).

To cover other types of masonry floors, such as slate or ceramic tiles, you must similarly ensure that the surface is dry, clean, and smooth. This is usually most easily done by removing the irregular surface, pouring a new slab, or installing a plywood subfloor over the masonry.

Preparing Wood Floors   Installing resilient flooring over a wood floor or subfloor entails the same preparation. If the subfloor is made up of individual boards, you have to install an underlayment, since the cracks between boards will inevitably show through. If the subfloor is plywood or composition board, it may suffice, but it must be reasonably smooth. If there are extensive cracks and holes, many nail heads, and some irregular areas, it will be easier to install your own underlayment.

The underlayment can be ¼-inch plywood or hardboard or thicker composition board, up to ¾ inch. Before laying the new sheets, make sure that the base is as sound and flat as possible. If there is any danger of moisture penetration (from a damp or poorly ventilated basement or crawl space), add plastic sheeting between the subfloor and the new underlayment as a moisture barrier.

The sheets of underlayment should be laid as tightly together as possible, but leave a small gap—⅛ to ¼ inch—around the edge of the room to allow for any expansion that changing temperature and humidity might cause. If there are baseboards, remove any quarter-round shoe molding (see page 23). Replace the shoe after the flooring is installed.

The sheets should be laid in a staggered pattern, so that no crosses between the junction of four sheets are formed. Position the edges over joists, where possible, so they can be nailed down. Don't nail too closely to the edges, however, since this risks damaging or deforming them, but be sure to nail securely and regularly, every six inches or so along every edge. The best nails to use are annular or cement-coated nails, and especially those with *sinker heads,* which are flat enough to be driven flush with the surface. Don't drive the nails below the surface, since this will cause a dimpling effect in the flooring. If you should accidentally drive any nail too deep, fill the hole!

If the underlayment is being installed over an existing finish floor, you must be prepared to cut off the bottom edge of any door that opens over the floor. The thickness of this cut would normally equal the thickness of the underlayment and the finish flooring combined, but it pays to wait until the job is done, then measure for the cut. Take the doors off while you're laying the new floor.

Preparing Existing Finish Floors   You can also lay resilient flooring directly over existing resilient flooring, providing it is level, defect-free, and still firmly adhered to its base. If it is not, you must either remove it, or cover it with another layer of underlayment.

If you decide to install new resilient flooring directly over existing resilient flooring, you cannot apply adhesive successfully if the old floor has been waxed or is not otherwise perfectly clean. Use the recommended solvent or cleaner in this situation. *Do not sand vinyl-asbestos flooring.* Much of it is made over an asbestos base, and sanding can release harmful asbestos fibers into the air.

Laying Sheet Flooring

When you plan the layout of a sheet floor, give some thought to how the pattern, if there is one, will fit best into the shape of the room. Large or repeated patterns look best if laid so that any trimming is equalized around the perimeter. If seams are necessary, do not place them in heavily trafficked areas. You may have to compromise at times; the most perfect layout may use the most material.

Laying Cushion Vinyl   Stretch cushion vinyls can be brought directly into the room and put into their layout position, leaving about 3 inches to extend up each wall (see Figure 7).

Then cut the edges of the sheet to fit, using a straightedge and a utility knife. Leave a small expansion gap along all edges; it will later be covered by the baseboard molding. The sequence of trimming is as follows:

Lay out the flooring; if there is a second section, it should overlap the first by a couple of inches, but the pattern should align perfectly (Figure 7). Make the first cuts at the two ends of the seam or seams, so that at the seam the flooring fits against the wall and lies flat on the floor. Using a straightedge and a sharp knife, cut through both layers simultaneously (see Replacing a Section of Sheet Flooring, page 31), having someone stand on the edges to prevent them from moving. Raise up the edges of the double-cut seam and apply adhesive to the floor below and press the two seam edges into place. Next trim the corners, punching through the flooring with a linoleum cutter and cutting upward into the rolled-up margin, thus allowing the flooring to lie flat on the floor.

After trimming them, attach the edges all around. Where they will be covered by molding, it is easiest to use a staple gun to secure the vinyl (see Figure 8). Where the edge will remain exposed, use adhesive. At doorways and passageways where there is no threshold or sill to be removed and reapplied over the flooring, special metal finishing strips, available from

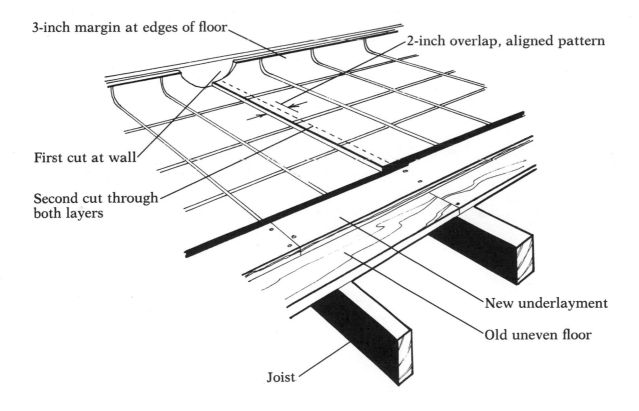

Figure 7   Laying sheet flooring

the flooring supplier, should be cut to length and nailed in place. Where the edge cannot be covered or otherwise concealed, such as around the bottom of shaped door architraves, try sliding the flooring under the obstruction, making a saw cut in the bottom of the molding if necessary.

An alternative for covering the edges is to use a vinyl base-molding all around the room. This is flexible strip material that is glued to the walls but not to the floors. Some types of strip are made with separate matching corners. For others, corners may be cut and mitered, or scored and stretched tightly to give a snug fit.

Laying Rotovinyl   The heavier sort of vinyl, known as *rotovinyl,* is best cut to size first, then installed. It will help to store the flooring in the room for which it is destined, at a temperature of at least 65 degrees, for 24 hours before installation.

Before cutting the vinyl to fit, make an exact floor plan. If the room is at all irregular, or if the corners are not exactly right angles, it is best to make a full-size pattern.

Making a Pattern   Use taped-together pieces of newspaper or lining, building, or butcher's paper to create a pattern sheet. First, take exact measurements from point to point in the

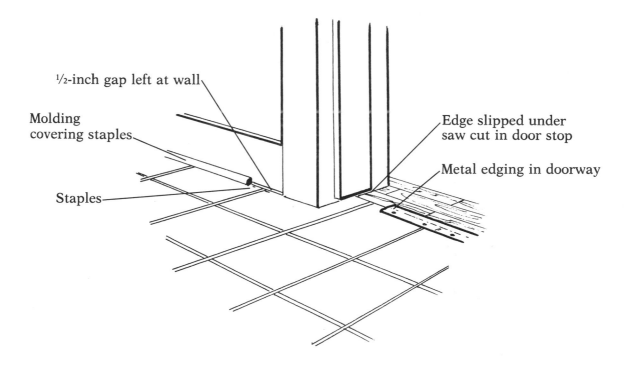

Figure 8   Edge treatments for sheet flooring

room and create a scaled floor plan with measurements written on it. When you're finished, double-check the written measurements. Then use a carpenter's square, straightedge, and tape measure to transfer the full measurements onto the pattern sheets. Cut out the pattern and check its fit. In a large room, do the room in halves. If only one area of the room is complicated, make an exact pattern of that area and transfer other room measurements directly onto the vinyl.

Cutting the Vinyl Unroll the flooring (with the face side up) in a clean and sufficiently large area and superimpose the floor plan or pattern on it. (Be sure to place the plan or pattern back on the flooring face.) Leaving a margin of about 3 inches all around, cut it to shape, using a stout utility knife or a linoleum knife and a straightedge for the straight sections, and heavy-duty scissors for any curved sections.

Depending on the manufacturer's recommendations, apply adhesive where required to the floor and, starting with the longest edge first, unroll the flooring and trim it, as described for stretch cushion vinyl. Rather than applying adhesive to the whole floor at once, it might prove easier to apply it one section at a time, unrolling the flooring as you go.

Laying Individual Tiles

Unlike sheet flooring, which is laid starting at one edge of the room, tiles are laid from the center of the room out. This is to ensure a better look and fit. (If you started at one wall, it is unlikely that the tiles would line up when you reached the opposite or adjacent walls.)

Start by finding the center of the room. This may be the exact geometric center—ascertained by measuring between opposite walls and marking the intersection of both halfway points—or the visual center, if the room is irregularly shaped. From this center point create two perpendicular lines, using a straightedge to draw the lines, then extending each line to the wall by snapping chalk lines. The lines may be parallel to the sides of the room or on the diagonal, if that is the pattern you choose, but they must be exactly perpendicular to each other. Check this by measuring 3 feet along one line (from the center point), and 4 feet along the adjacent perpendicular line. If they are truly perpendicular (at right angles), the distance between the 3-foot mark and the 4-foot mark will be exactly 5 feet (see Figure 9). If not, adjust the lines until this is true.

Lay a row of tiles (without adhesive) along each line, adjusting them so that the gap between the last tile and the wall is the same at both ends of the row. The gap will be filled with cut tiles, which will form equal borders at opposite ends of the room. Do this in both directions; you'll probably have to remark the perpendicular lines—these are your final guidelines.

If the tiles are to be laid in adhesive, apply only as much as you can tile over before the adhesive sets. Read the instructions on the adhesive to know how long you have and what application tools are recommended. *Do not obscure your lines with the adhesive!* Now you are ready to begin laying the tiles, which you should do in a pyramidal fashion (Figure 9),

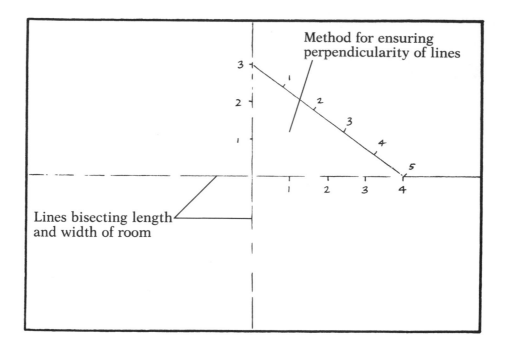

Method for ensuring
perpendicularity of lines

Lines bisecting length
and width of room

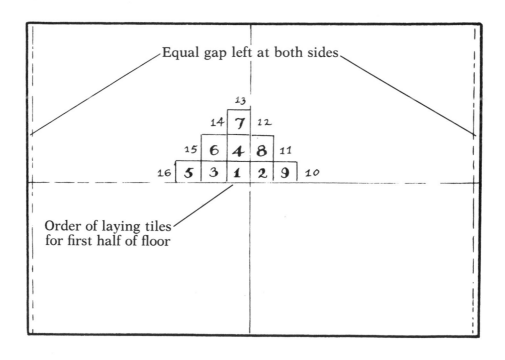

Equal gap left at both sides

Order of laying tiles
for first half of floor

Figure 9   Tile layout

covering first one quarter of the room and then the others, starting each time from your center point.

By doing this, each successive tile has two edges of previously laid tile to butt up against, giving you the least possibility of misalignment. Leave the spaces of less than one tile's width around the perimeter of the room until the very end. Fitting these smaller pieces is done by either the one-tile or two-tile method. You can lay a full tile over the gap, then mark and cut it along the line where it overlaps the previously laid tile. Or you can put one tile directly on top of the last tile in the row, put a second tile over it and butt the second tile to the wall, then mark and cut the first tile at the point where the second tile overlaps it. The piece remaining will fill the gap. Make patterns to fit tiles around pipes and other irregularities.

4

CERAMIC TILE
AND MASONRY FLOORS

ASSESSING THE CONDITION

Taking stock of ceramic tile flooring involves finding out what kind you are faced with. There are various types, often classified differently according to size or material. However, note that ceramic tiles are either *glazed* or *unglazed*.

Glazed means that a hard surface, often colored, has been applied between the first firing and the final firing in making the tile. There are, of course, different glazes, ranging from high gloss to dull, pebbly finishes. Tiles are made in innumerable shapes and sizes, from individual 1-by-1-inch squares to sheets of tiles held together by some form of backing.

Unglazed tiles are naturally dull, although surface finishes such as sealers and waxes can add a shine. The color of these tiles is the result of the color of the clay from which they have been made, to which a pigment may have been added. Unglazed tiles include:

- *Quarry tiles,* which come in natural colors (reds, grays, and browns) and in various sizes (generally between $3/8$ and $7/8$ inch thick) and which are more or less water-resistant
- *Pavers,* which are more rugged and more water-resistant (some may even be glazed), and which are generally available in standard sizes: 4 by 4 inches, 6 by 6 inches, and 4 by 8 inches
- *Patio tiles,* less uniform than quarry *tiles* or pavers, best suited for indoor or warm-weather outdoor use (they may shatter if they freeze)

- *Mexican* or so-called *primitive tiles* of terra-cotta, generally larger and more irregular than the other types, which need to be sealed to prevent their surfaces from powdering

Masonry refers to natural materials such as clay, stone, and rock. Strictly speaking, that would include ceramic tiles, but usually the term *masonry* includes only brick, terrazzo, flagstone, and other stone types such as granite, marble, and limestone.

Fundamental to the condition of any interior ceramic tile or masonry floor is the state of the floor on which it is laid. Any deterioration of the underlayment, the subfloor, or its supporting structure will surely threaten the integrity of the finish floor. Before attempting any repairs—or even coming to any decisions—make sure that all is well below. Refer to the section on assessing the condition of floors in chapter 2 (page 12), and make any necessary repairs below before attempting to repair or replace any tile or masonry finish floor.

Cosmetic problems such as dirt, stains, and cracked and chipped elements in the system (such as a loose or missing tile) usually can be repaired easily. If there is more general damage, such as excessive cracking or powdering, or if there are large areas that are damaged and missing, removal and replacement are in order.

REPAIRING AN EXISTING FLOOR

Cleaning a Tile Floor

The repair most commonly called for on ceramic tile is cleaning. You don't need to be handy to clean ceramic tile, but you must be careful, patient, and persistent. Hard accretions on the surface must be scraped or chipped away, with special care taken not to damage any glaze. Hardened tile adhesive may be removed with varying degrees of success with paint thinner. Stains and discolorations are more common with unglazed than with glazed tiles. Heavy-duty household detergents will deal with most of these.

Some problems require stronger treatment. Lime deposits, which may occur on bathroom tiles if you live in a hard-water area, can be dissolved by soaking with vinegar. Rust stains will respond to a washing with a solution of 5 percent oxalic acid, or a solution of 10 percent hydrochloric acid. Copper stains—green areas around pipes that disappear into the floor—can also be removed with a 5 percent solution of oxalic acid, though you should try a strong solution of soap and ammonia first. Be extremely careful when using these acid solutions. Wear safety goggles, long sleeves, and rubber gloves, and if any of the solution comes in contact with your skin flush the area immediately with plenty of water.

Grout—the material in the seams between tiles—often becomes dirty, discolored, or mildewed. Scrubbing with a household bleach or tile cleanser usually produces results, but before using bleach on colored grout, which may react with the tile or grout, test a small, inconspicuous area first. Sometimes a toothbrush is handier than a regular scrub brush to get into corners and behind or between pipes.

Cleaning a Masonry Floor

Cleaning is generally less of a problem with masonry floors than with ceramic tile floors since discoloration is often accepted as part of the aging patina. Vigorous scrubbing with regular household cleanser, however, can remove a surprising number of spots and stains. For a more thorough cleaning, you may have to resort to stronger commercial washes. Although they are very effective, use them only after having tried regular detergents. Then be sure to wear protective clothing—safety goggles, face mask, long sleeves, and rubber gloves—and follow all the precautions on the label. Flush off with water any solution that comes in contact with your skin. Avoid using muriatic acid. It is an extremely hazardous substance. Not only are its vapors harmful, but it can burn your skin and cause permanent damage to your eyes.

Oil and grease spills can be treated by first soaking up the excess with some absorbent material, such as cat litter, then scrubbing with detergent. Other stains may require the use of cleaning poultices. These pastelike substances contain both a chemical solvent that penetrates and dissolves the stain and an absorbent material that soaks it up. After using it, you can simply brush away the whole mess.

Efflorescence is a stain that has nothing to do with dirt. It's a whitish powder that is left on the surface of brickwork when salts are leached out by exposure to moisture. Efflorescence may also be removed with concrete-etch. When the area is clean, it must be flushed well with clean water.

Cleaning with any of the solutions mentioned can also affect the masonry floor's surface finish—such as a sealer or polish—and this may in turn have to be touched up or redone.

When the Floor Is Damaged

Before dealing with any visible damage—such as cracks in the floor or loose tiles or pieces of masonry—remember our caution: Make sure the damage has not been caused by underlying problems. If that is the case, no repair will hold for long. Note particularly that a continuous crack through two or more tiles is often a sign that something is amiss below. Nevertheless, things do get dropped on floors, and from time to time it may be necessary to repair small areas. Making repairs on ceramic tile and masonry floors is not particularly difficult. Work carefully to avoid breakage.

When dealing with stone, brick, or commonly available tile, the availability of matching material represents no problem. But since ceramic tile comes in literally hundreds of sizes and colors, it may not be possible to secure a perfectly matched tile or tiles. If you can't, you'll have to decide whether the closest match you can find will look better than simply leaving the crack, or filling the gap with mortar. Repairing ceramic tile is not difficult, but it does require a light, deft touch.

Removing Damaged Material When chipping out a section of a ceramic tile or masonry floor, always wear protective glasses. To minimize the risk of cracking further areas, apply masking tape around the edges of the area in which you're working.

Even if you have several tiles to remove, remove them one at a time with a chisel and hammer. If you're trying to remove a complete tile, chip out the surrounding grout in this manner: Make a small vertical hole in each seam of grout and use these holes as the starting points. Work out from the holes, holding the chisel at an angle of about 45 degrees. Use many light taps with the hammer, and avoid heavy blows.

If you are unable to remove a whole tile (part of it may now be concealed under a built-in cabinet or bathroom fixture, for example), use a glass cutter to score a line as close to the obstruction as possible. Then, score crisscrossed diagonal lines across the tile and, using a masonry bit, drill a ¼-inch hole where the lines cross. It should now be relatively easy to break the tile into pieces by chiseling outward from the center hole along the scored lines. If you are removing marble or slate, it is safer first to drill a series of holes about ½ inch in diameter and about ½ inch apart. When the tile has been removed, scrape, chip, or wire-brush as much of the old grout and adhesive away as possible.

Inserting New Material Cut a new replacement to size (see page 48). Apply whatever adhesive is appropriate for the material, the subfloor, and the condition—such as dampness (see page 45). Install the new material, using shims to keep the gaps around the edge consistent—woodstrips, nails, or popsicle sticks are useful.

The final job in the repair is to replace the grout. This should be done following the method described in Grouting (page 48)—though, since the area is small, you may find it simpler (if the color of the old grout and the color of the new grout are similar) to use premixed silicone grout. It comes in tubes, is ready to be squeezed out, adheres well to both old and new tile, and cures quickly.

INSTALLING CERAMIC TILE

Selecting Materials

Choosing which kind of tile to use may well be the most confusing part of the job. The different types were discussed at the beginning of this chapter. One dealer cannot possibly carry all of them, so visit several dealers. Go armed with a floor plan that is measured and drawn to scale on graph paper, and be prepared to order more tile than you think you'll need. There will almost certainly be some breakage if you have any cutting or trimming to do, and it's a good idea to store some extra tiles in case future repairs are needed. Installing ceramic tile is somewhat finicky work, particularly cutting tiles and grouting. You'll need to rent a tile cutter and tile nippers, and to construct a tile leveler, described in a later section. You'll need to purchase special tools, a toothed adhesive trowel, a rubber-based float, and a striking tool.

Choosing an Adhesive   Tiles can be laid in two different ways, depending partly on the floor they are to cover and partly on the use to which they are to be put. The traditional method is to set the tiles with a portland cement paste on a bed of mortar about 1 inch thick. This is known as the *thick-set* method. It is ideal for use over less than perfectly smooth concrete slabs and when you are laying uneven tiles, such as Mexican tiles, which vary in thickness. But the thick-set method requires considerable skill, and is best left to professionals.

The alternative is the *thin-set* method, in which the tiles are laid directly over the base with a bond coat of mortar or adhesive (see Figure 10). Using adhesive is relatively easy for the do-it-yourselfer, providing the correct one is used. It should be determined by the kind of tile you are using and the kind of subfloor on which it is to be installed. There are three basic types of thin-set adhesives: mastics, mortars, and epoxies. As always, when working with adhesives, cross ventilate the room with open windows and doors and a window fan set to exhaust air.

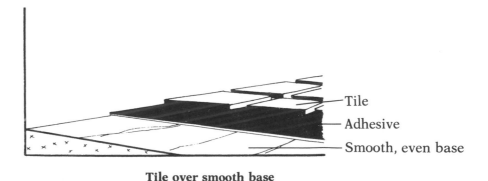

Tile
Adhesive
Smooth, even base

Tile over smooth base

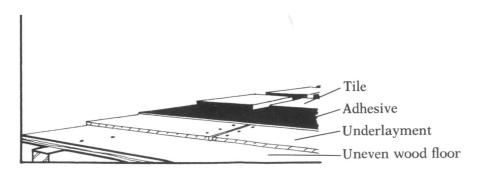

Tile
Adhesive
Underlayment
Uneven wood floor

Tile over underlayment

Figure 10   Thin-set tile-laying methods

Mastics are paste adhesives and come in two types. The first is water resistant since it is made with a chemical solvent. Although that makes it ideal for use in potentially damp areas, such as bathrooms and basements, it is inflammable and toxic until it sets. When you use this type you must wear gloves and a protective mask. And you must take extra care to provide adequate cross ventilation. The second type, which is formulated with a latex base, is much more benign to use. Although that makes for safer application and easier cleanup, the latex type is really only good for moisture-free areas. Both types are fine for use over masonry, concrete slabs, or even plywood subfloors, providing the surface is firm and level.

Mortar adhesives are cement based, which means they can easily be cleaned up and are not toxic. They are usually limited to installation over existing masonry or over concrete. Their big advantage over traditional mortar beds is that the tiles being laid need not be presoaked. As a result, they are known as *dry-set mortars.* One type of dry-set mortar has to be mixed with liquid latex before use. That makes the resultant paste much easier to trowel and use, and also increases its water resistance. With all dry-set mortars, it's important to follow the package directions for mixing. There are certain critical waiting periods, which bring the mortar to just the right consistency.

Epoxy-based adhesives are the strongest and the most universally applicable—usable over concrete, plywood, and even in damp areas—but the most expensive. They are also toxic; wear rubber gloves when you apply it, and cross ventilate the room with a window fan. Temperature is important to epoxies, both when they're being mixed and when they're curing (hardening). Once again, it is imperative to follow all package directions exactly.

Choosing a Grout

In addition to the actual tiles and the adhesive, you'll have to decide what type of grout to use. There are cement-based, mastic, epoxy, and silicone rubber grouts.

All *cement-based grouts,* the type most commonly used, are relatively easy to use, with good water retention and enough flexibility for most tile applications. Dry-set cement grout is sold as a premixed powder, which has only to be mixed with water before using. It is known as dry-set because the tiles being laid do not have to be presoaked.

Latex cement grout, like latex cement mortar (described above), has to be mixed with liquid latex prior to use. It is more water resistant. *Portland cement grout,* still another variety, requires the tiles to be presoaked, and is thus best suited for use with quarry tiles and pavers.

Mastic grouts are sold in paste form. They tend to be more flexible and more stain resistant than cement-based grouts.

Epoxy grouts are the most water resistant and create the strongest bond, but they are difficult to work with. Temperature is a crucial element when mixing and using them, and cleanup can be very difficult if postponed too long. Though epoxy grouts are among the most expensive, their cost is often justified by the excellent, long-lasting job they do in high-traffic areas. A further advantage of epoxy grouts is their wide range of colors.

Silicone-rubber grout comes in ready-to-squeeze tubes, ideal for applying around bath-tubs and sinks. It is primarily a repair grout, to use when replacing damaged floor tiles or sealing a space between tiles and some other material (such as wood). Silicone-rubber grout is not really appropriate for general use in laying floor tiles because it's too soft when it sets.

Even with these comments in mind, the choice of a grout may still seem confusing since there are so many variables—different tiles, different bases on which they are to be laid, and different adhesives to be used. Very often the best course to follow will be to read labels carefully and match all the ingredients to the particular project.

Preparing the Subfloor

Ceramic tiles require a perfectly firm, level, and smooth subfloor. The underlying structure must obviously be in good condition (see page 33). The best of all possible bases is a concrete slab, but it must be completely cured, absolutely dry, and clean. If not, it can impair whatever adhesive you use. Certain adhesives call for a sealer to be applied to the concrete first—follow the label directions.

Directions for successful application of ceramic tiles over a wood subfloor are similar: The surface must be firm, level, dry, and clean. Sufficient firmness—a floor that does not give when you walk on it—may usually be achieved by installing an additional layer of ½-inch plywood. If this is not enough, then the floor needs more support from below (see chapter 2). If you are working directly over a board subfloor, it will almost certainly be necessary to install an underlayment of plywood first, in order to provide a continuous surface. Nail the plywood down securely, at least every 6 inches along the edges. No nail heads should protrude from the surface of the plywood. An existing wood finish floor should be similarly checked for any nail heads or irregularities and given a light sanding to remove any finish that might impair adhesion of the new adhesive.

Other floor surfaces—such as resilient tiles and other tile and masonry surfaces—can also serve as a base for new ceramic tiles, if they are clean, level, and themselves firmly attached to whatever is beneath. Bear in mind, however, that installing tiles over certain surfaces requires the use of specific adhesives; for example, epoxy adhesives are generally needed over resilient flooring. Your tile supplier should be able to recommend the appropriate adhesive if he knows what tiles you are using and where they are to be installed.

If you are installing tiles over a finished floor, it will be necessary first to remove baseboard shoe molding, pipe escutcheons, and floor grilles. Doors that open into the room may have to be adjusted to accommodate the increased floor height. This is occasionally difficult enough that it may pay to remove the existing floor before applying a new layer—weigh the amount of work involved in each case before you begin. In either case, remove the doors before tiling the floor.

Laying the Tiles: Thin-Set Method

Planning the Tile Layout   Since few rooms are perfect rectangles, and since you may want to use contrasting border tiles, or even lay a pattern, it is usually easier if you don't start right along a wall. One way to lay out the pattern for the room is to start in the center, as explained in chapter 3 (see page 37).

An alternative is to establish a straight line approximately one tile's width (including the width of the grout) away from one wall and a second line at a right angle along an adjacent wall. Place tiles in the corners of these walls and snap chalk lines between them. You can ensure a perfect right angle by using the 3-4-5 method shown in Figure 9. Once you have established an absolutely right angle at one corner, adjust the pattern for minimum cutting. Tack strips of batten along the chalk lines so you have a straight edge against which to lay the first row of tiles. You then lay tiles from the square corner, and must cut and fit tiles along the two "unsquared" walls. This method is most effective in smaller rooms.

Before applying any adhesive it's a good idea to lay out a row of tiles dry first, spacing them correctly, in order to see how they fit. The correct spacing for the grout can be established by using thin strips of wood or plastic spacers (usually available from tile suppliers). If the tiles you've laid out don't make a complete row, it's best to adjust the row so the extra space is left at each end, to be filled after the central portion is completed. The cutting of tiles is thus confined to the edges (which are often not square and likely to require trimmed tiles in any case).

You may want to consider starting the tiling from the point at which the floor will be first seen—from the entrance to a bathroom, for example—leaving any interruptions in the pattern to fall in less visible areas. If one section of the floor will be the focal point, it may make better aesthetic sense to tile this area first. In short, bear both aesthetic and practical considerations in mind when deciding how the patterns should fall and where to start tiling.

Starting to Tile   Once you've decided on the actual tile layout, you're ready to begin laying. Ventilate the room and don't apply too much adhesive all at once. As you become more adept, you'll be able to cover larger areas, but take your time at first, positioning the tiles carefully and leaving the correct gap between tiles. At the start you'll have the battens as guidelines. Thereafter, use spacers and snap chalk lines to guide your progress. (Beginners should snap a new chalk line every row; as you gain experience snap a line every three or four rows.) When snapping new chalk lines, always measure from the original point, not from the last line of tiles laid. If you measure from the last row, small errors can compound to become gross distortions.

Apply sufficient adhesive, but don't apply so much adhesive that it squeezes up between tiles, leaving too little space for the grout. Set the tiles in place with a twisting motion to ensure complete contact. To make sure that the tiles are set firmly and are level with each other, use a piece of plywood—12 to 18 inches square is ideal—covered tightly with

47

carpeting. Place the carpet-covered face over a section of tiles and hammer gently but firmly over the back surface of the wood.

Cutting Tiles   Sooner or later you'll have to cut tiles to fit. Rent a *tile cutter* from the tile supplier, a hardware store, or tool rental center. The tile cutter is actually a glass cutter mounted on a frame in which a tile is held firmly and squarely while it's scored with the glass cutter. Make sure the little wheel that does the scoring turns easily and is sharp.

The procedure is simple. Mark the face of the tile where it is to be cut, place it in the cutter, score it along the line, and remove it. Rest the tile, face side up, with the scored line positioned exactly over a rod or pencil and press down on both sides. It should snap easily and clearly at the line. (Some tile cutters use a lever to break the tile along the scored line.)

For cuts that aren't straight, use a pair of *tile nippers* (common pliers will also work, but the spring-loaded tile nippers are better). Score the section to be removed in a grid pattern and break out pieces with the tile nippers until the curved line is obtained. With some tiles, it's possible to smooth out a cut with a fine file.

Cutting a tile around a pipe is most neatly done if the pipe can pass through in the middle of a tile. Put a piece of masking tape over the point where you want to drill the hole, and, with a masonry bit, drill the appropriate hole. Then score a line across the tile so it bisects the hole. Snap the tile in two and install each part on either side of the pipe. The center seam will be barely noticeable.

Cleaning Up

Once the tiles are laid, you have a short period in which to make minor adjustments. Be sure to remove the spacers before the adhesive sets. Clean up as you go—epoxy adhesive is especially hard to remove once it has started to set. Unless it's absolutely necessary, don't walk on the tiles before the adhesive is completely cured. That varies with temperature, the weather, and the kind of adhesive you are using (it can take less than a day or up to two days). If you must walk on the tiles, place a length of board or plywood over them before you do. That spreads out your weight and prevents the tiles from moving. It also avoids damaging the tile edges, which are especially vulnerable until the floor is grouted.

Grouting

Be sure you have the appropriate grout for the kind of tiles you have used. No matter what grout you use for the main area, you may need a silicone grout where tiles butt up against other materials, such as wood baseboards or a bathtub. Choosing the right grout is especially important with unglazed tiles, since certain grouts may stain them.

Applying the Grout   In general, most grouts are applied in a similar fashion. The main point is to ensure that the grout completely fills all the seams, with no hidden air pockets.

The usual method is to spread the prepared grout liberally over the surface of the tiles with a *rubber-based float* used in a swirling motion. Since many grouts are irritants, wear gloves. Before you start, ventilate the area and make sure you have any necessary solvent handy (this may be plain water).

As soon as you have completely filled all the seams in any one section, wipe off the excess grout, leaving the tiles as clean as possible. A damp sponge is usually the best tool at this stage; keep it clean by constant rinsing and squeezing in a bucket of clean water. Change the rinse water frequently. By the time you've grouted the whole floor, the first sections will usually have dried, leaving a thin haze on the surface. Polish this haze away with clean, soft rags.

Now you have to "strike" the joints. This means simply running a *striking tool* along the grout to smooth it down evenly in the seams. This can be a professional striking tool or any smooth object narrow enough to fit in the seams, such as the end of a pencil or a toothbrush handle.

Curing Most grouts take a surprisingly long time to cure (harden completely), so be patient. You may even have to keep cement-type grouts damp during this period by covering the floor with plastic. Some grouts may also require sealing after they have cured; check the instructions on the container. Avoid walking on the floor until the grout has cured, or lay down pieces of board or plywood on which to step. After the grout has cured, replace any grilles, covers, or shoe molding.

INSTALLING OTHER MASONRY FLOORS

Installing a stone or brick floor is considerably heavier work than installing ceramic tile, but the technique is no more difficult. You'll need a few brick-working tools: a soft-headed steel hammer, a brick set, a brick hammer, a trowel, and a pointing (small) trowel. If you use mortar, you'll need equipment in which to mix the mortar: a mortar box and masonry hoe or a small power mixer, which can be rented from a tool-rental company.

Brick or stone is most commonly found in entryways and around fireplaces, but these materials are sometimes appropriate for other areas, such as kitchens and bathrooms. Unless there is an existing concrete slab, make absolutely certain that the floor structure is capable of supporting the weight of the material you plan to use. Obtain the advice of an architect or building engineer rather than guessing.

Brick Floors

Bricks designed for use as flooring are called *pavers,* which are roughly the same size as regular bricks, or *splits,* which are only half as thick. Using splits lightens the load the floor will have to carry, but still requires a stronger structure than do other finish floors. Bricks can be laid either with or without mortar. The latter is perhaps more handsome and somewhat longer lasting, but the former is very much easier and far quicker.

Laying Mortarless Brick   Start by checking that the floor is strong enough to support the load and is level. To prepare the floor, lay down a covering of *felt building paper* (see Figure 11). Felt building paper is made in different thicknesses, or weights; 15-pound felt paper laid so that the edges butt up against each other tightly is sufficient. Overlapping is unnecessary. Put down a second layer at right angles to the first to prevent drafts or debris from passing through.

Choose one of the many patterns in which bricks can be laid; Figure 12 shows a few of the possibilities. One of the advantages of pavers is that two widths exactly match one length. (Wall bricks have slightly different proportions to allow for the mortar joints.) Start laying your pattern from the straightest part of the area to be covered, using a chalk line to establish a reference point, if necessary. If the disparity between the wall line and the chalk line is great, lay the bricks along the chalk line first, then cut bricks to fit the remaining gap(s).

Cutting bricks requires a *soft-headed steel hammer* (one that won't chip on impact) and a *broad-bladed cold chisel* (also known as a *brick set*). Mark the brick where it is to be cut and score it by drawing the point of the chisel along the line. Make the cut by tapping the chisel into the scored line. Different bricks have different consistencies, and with a little experience you'll know how much force you need to make the cut without shattering the brick. You can trim lumps protruding at the cut edge with a *brick hammer*, one end of which is chisel-shaped for just this purpose.

When the first *course* (line of bricks) is laid and any short bricks cut and installed, proceed with the next course. As when laying tiles, check every three or four courses (with a snapped chalk line measured from the original starting point) that all is still correctly lined up. How this is done will vary, of course, depending on the pattern you've chosen. Although you're trying to butt each brick up against its neighbor as tightly as possible, you may have to make occasional slight allowances to achieve an overall regularity.

When the entire area has been laid, sweep very fine mortar sand into the joints. This can be a little frustrating at first. The initial sweeping, which appears to have left sand-filled joints flush with the bricks' surfaces, will inevitably require repeating, as the sand works its way down. After a few days, if the sand-filled joints are still flush, seal the entire surface with a masonry sealer and the job will be complete.

Laying Bricks in Mortar   Laying bricks in mortar requires either a very substantial subfloor or, better still, a concrete slab. If the slab is clean and reasonably level, the mortar bed in which the bricks are laid will compensate for minor irregularities (see Figure 11). The layout procedure is the same as for laying a mortarless brick floor except for one important point: Since it won't be possible to walk on the freshly laid bricks until the mortar has cured, you must plan the layout so that everything is cut and laid as you go. And you must work

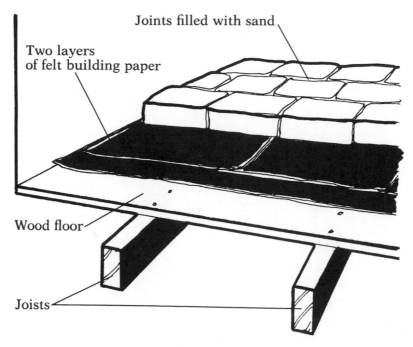

Joints filled with sand

Two layers
of felt building paper

Wood floor

Joists

Mortarless brick floor

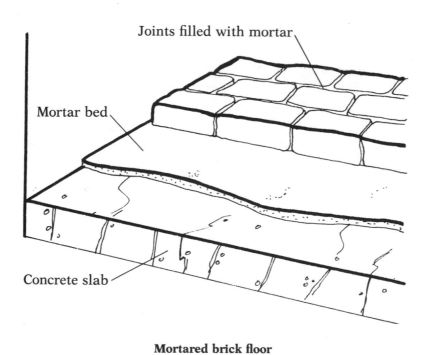

Joints filled with mortar

Mortar bed

Concrete slab

Mortared brick floor

Figure 11   Brick paving methods

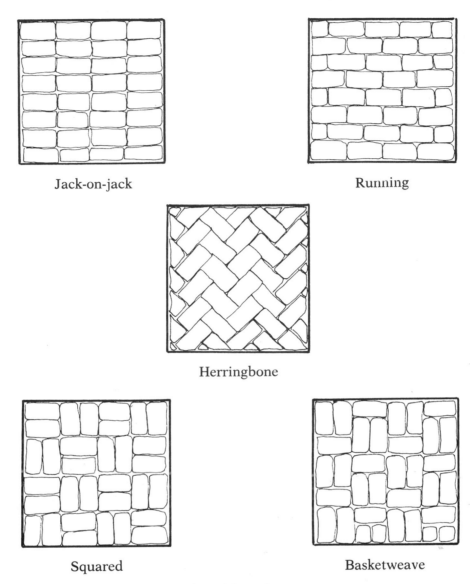

Jack-on-jack   Running

Herringbone

Squared   Basketweave

Figure 12   Brick paving patterns

toward an exit! Laying bricks in mortar also requires considerable skill, and should not be undertaken by a novice do-it-yourselfer.

The actual mortar bed need only be about 1 inch thick, but the consistency of the mortar has to be just right. Mix it in amounts that will be used within an hour. Try to achieve a consistency that is lump-free, easy to spread, but not so loose (wet) that a good dollop will slump when trowelled out of the mixing tray. The basic recipe is three parts sand, one part portland cement, and one-quarter part lime. The exact proportions will vary, depending mainly on the condition of the sand. Ideally, you should be working with perfectly dry, clean, graded sand, but this is not always possible. Aim for the right consistency rather than sticking exactly to the recipe. Premixed mortar is very convenient to use, but it is more expensive.

The bricks must be wet through, as they are laid. Dry bricks will absorb water from the mortar, weakening it and threatening the integrity of the bond. Keep a steady supply of bricks soaking in a bucket of water—it only takes a couple of minutes for the bricks to become wet enough, but it doesn't matter if they soak longer.

Apply as even a layer of mortar as possible (about 1 inch thick) to a small area at a time, and set the bricks in the bed according to the pattern you've chosen. Use a stretched line to ensure you're laying a straight row, and check constantly with a mason's level to ensure surface flatness, tapping the bricks down with the end of the trowel handle when necessary.

The bricks should be spaced as regularly as possible with joints about ¼ inch wide. Use wood spacers to help you maintain the correct joint width, but you do have leeway to make small adjustments in order to maintain the regularity of the pattern. Remember to complete each section as you go, since you cannot step on the bricks to fill in odd spots later.

The hardest part is keeping everything clean and neat. You'll inevitably drop bits of mortar onto the bricks, however, so be prepared to clean them up with a coarse sponge or a piece of sacking soaked in water.

Twenty-four hours later you can *point* the bricks—fill in the joints. Use a board on which you can kneel to distribute your weight over the bricks. Point the joints very carefully with mortar of the same consistency, but use a small pointing trowel to achieve as neat and clean a job as possible. Once again, wipe up any spills, spatters, or drops as you go. When the mortar in the joints has dried so you can smooth the surface without disturbing it further, do so with a striking tool as evenly as possible. Once again, check for smears or drops of mortar—this is your last chance. Finally, sprinkle the entire surface lightly with water and cover it with a sheet of plastic so the entire floor will cure at the right speed. Depending on the local humidity, leave the plastic in place for one to two days. Although you can now safely walk on the floor, take care to keep it clean since it is not yet sealed. Sealing should be done only when the floor is completely dry (trapped moisture can damage the sealer). To be safe, wait for two or three weeks before sealing.

Stone Floors

Although there are countless types of stone that may be used as flooring, they can be split into two main groups: *cut-and-finished stone,* and *rough-hewn stone.* The first group includes *terrazzo* (made of marble chips embedded in cement, polished smooth) and pieces of other stones such as granite, limestone, sandstone, or slate cut into uniform thickness. These materials, though heavier, may be installed in the same manner as ceramic tile, using appropriate mortars and grouts.

The second type, rough-hewn stone, is installed similarly to mortared brick floors, and the same cautions apply in regard to weight. Since most stone is less absorbent than brick, it is not necessary to soak the stone. On the other hand, you're likely to have a lot more cutting to do than with brick. Careful layout can minimize this, but you will inevitably have to cut large pieces. Some stone, such as slate and dressed sandstone, may be cut with a standard circular saw fitted with a masonry blade. This is good for straight edges, such as along the walls of a room, but if you are laying randomly shaped pieces, a sawn edge will look out of place.

Different types of stone cut differently. Careful scoring and gentle but persistent tapping with a soft-headed steel hammer on a cold chisel, with the stone supported in such a way that the waste is off the ground and can fall, will usually produce the desired shape. Experience will teach you best how each kind of stone tends to fracture. *Sedimentary stone,* such as limestone, possesses a regular structure called its *grain,* along which it is predisposed to break. *Igneous stone,* such as granite, tends to fracture more irregularly. Terrazzo is difficult to work with and installation is best left to professionals.

MAINTENANCE

Ceramic Tile Floors

One of the major advantages of glazed ceramic tile is that the surface is well protected and needs only to be washed occasionally. The most frequent problem occurs with the grout, which can become dirty, stained, or mildewed. Grout that has deteriorated badly can usually be dealt with only by retiling the floor. Short of that, however, there are many products available to remove mildew and clean the surface of grout. Beware of using household bleach on colored grout—it may react with the grout and change its color. Try it on an inconspicuous spot first.

The surface of unglazed tile must be sealed so that dirt is not ground into it. However, there are many products available to seal the surface and impart either a high (glossy) or subdued (matte) shine. Unglazed tile may be waxed on a regular basis with special tile wax, which both protects it and improves its color.

Other Masonry Floors

Apart from being treated with the correct sealer, and a possible occasional waxing, most masonry floors require very little maintenance. Sweeping, mopping, and repolishing are all that need be or, indeed, can be done. That is part of the reward for the extra labor involved in installing them.

5

CARPETED FLOORS

TYPES OF CARPETING

Selecting Materials

This chapter is concerned mainly with carpeting as a complete floor covering, in the form commonly known as *wall-to-wall carpeting*. Nevertheless, much of the discussion on repair and care is also applicable to individual carpets and rugs.

Carpeting, one of the most luxurious of all floor coverings, has been popular since the eleventh century, when Oriental carpets were first introduced to the Western world by returning Crusaders. Hand-woven, *pure wool* carpeting is still considered to be the best, but carpeting is now made of other materials that cost far less. These less expensive materials have made it a true alternative form of floor covering, providing an almost limitless choice of color and design at reasonable cost.

The Fiber Lineup   The new materials, sometimes combined with each other, sometimes with regular wool or even cotton, have different characteristics than wool, apart from the cost. *Acrylic,* for example, although no longer as common as it once was, is especially long wearing. *Nylon,* which no longer has a serious static-electricity problem, is especially useful because of its stain resistance. *Polyesters* feel very luxurious and resist abrasion well, but some are not so easy to keep clean; and *polypropylene,* because it fades little and is not absorbent, is commonly used in homes both as regular and as indoor-outdoor carpeting.

Conventional versus Cushion-backed Carpet   Aside from individual carpets, and carpeting in squares designed to be laid like resilient tiles, carpeting falls into two categories: *conventional* and *cushion-backed*. Unlike traditional woven carpets, these two types are made by the tufting process, in which the fibers are stitched into a backing material, anchored by a coat of latex, then given a secondary backing of jute or polypropylene.

Conventional carpeting is made without an integral rubber backing and is the type designed to be stretched permanently in place over a separate underlayment of padding. Cushion-backed carpeting is manufactured with the padding as part of the carpet; it may be simply laid in place or installed with adhesive. Cushion-backed carpeting is usually much cheaper than conventional carpeting and is much easier to install. It does not last as long, however, and it has one further disadvantage: If it has been laid with adhesive, it is usually ruined when it is taken up—the adhesive cannot easily be removed from either the floor or the backing.

Types of Carpet Construction   Both types of carpeting can be constructed in various ways to give different effects, and it will help to understand the structure when you're trying to determine what type you're looking at.

For one thing, carpeting differs in the way the fibers are stitched into the backing. If they are left in one continuous thread, they form what is known as a *continuous-pile* or *level-loop* carpet. Such a carpet is very durable and easy to maintain, although a variation known as *multilevel-loop,* which hides dirt better, is harder to vacuum. If the top of the loops are sheared off, *cut-pile* carpet is formed. When woven very tightly, this type is also known as *plush* carpeting—it has a very smooth look that produces shadows (from the grain) after vacuuming. *Shag* carpeting is less dense and often has fibers of irregular length. It is less expensive and has a casual look, but it may not last as long as the more expensive types.

Sometimes part of the carpet is cut at different heights; this produces a design known as a *sculptured* carpet. A sculptured carpet design may or may not use the sculpted effect in conjunction with different colors. An additional element in carpet construction is the tightness with which the yarn is wound, known as the *twist*. Variations in the twist produce different effects seen in types known as *Saxony* and *frieze*.

Buying New Carpeting

To buy the right carpet for any particular location you must choose not only from a vast array of colors and patterns but also from the many types described above. Bear in mind the job the carpet will have to do. Its projected use will determine what material is most suitable; your budget will determine the level of quality.

Choosing Type and Color   For formal but little used areas, plushes and Saxonies give the best effect. But they are not as suitable for heavily used areas like living rooms since they show footprints and are not as easily cleaned or as hard-wearing as textured varieties like frieze, multilevel-loop, or cut-pile. Use will also play a role in choosing color. Light, easily soiled colors may be more appropriate in bedroom areas where shoes are less likely to be worn than in dining or family rooms.

Places where moisture is a factor, such as kitchens and bathrooms, require synthetic carpeting with moisture- and mildew-resistant backings. Washable carpet is especially appropriate in bathrooms. Entryways, stairs, and halls are best carpeted with low level-loop or dense cut-pile carpets. These wear best and look better longer, especially if chosen in darker, solid colors.

Having decided on the type of material, your next decision is the color and pattern. Carpeting is one of the dominating elements in the decorating scheme of any room. When you go carpet shopping, take along as many samples of your home's existing color scheme as you conveniently can. Since swatches of carpet in sample books may differ slightly from individual lots of carpeting, ask if you can see a sample of the actual stock being sold in a particular color.

Planning the Carpeting Layout   Go armed with as exact a floor plan as is possible. A scaled drawing on grid paper is often the easiest way to do this. Most carpeting is manufactured in 12-foot widths. The way you lay out the sections of carpeting needed for a large space will greatly affect the cost.

When you're trying to plan the most efficient layout, keep the following points in mind:

- A pattern may well dictate how the carpet is to be laid; a seam between sections, if necessary, will be less obvious if aligned with the major light source rather than at right angles to it (when a more noticeable shadow will be cast).
- Try to avoid seams in the most heavily trafficked part of the room. Nevertheless, if an extra seam is acceptable, you can often save a considerable amount of money by using pieces rather than buying another 12-foot width.
- Since tufted carpet is rolled up when it is made, the pile assumes a slope direction that it maintains for life. Since carpeting usually looks best when viewed from the direction toward which the pile lies, plan the layout so the pile points toward the place from which the carpet will be first, or chiefly, seen. This usually means orienting the carpet so that the pile points toward the door through which the room will be entered.
- No matter how you plan the layout, allow for a little extra to cut and fit, at least 3 inches all around, and possibly a little more for future repairs.

Judging Carpet Quality   One of the surest measurements of quality is the *face weight*, the measurement of how high the pile is and how many tufts of yarn there are in each inch of carpeting. Although the face weight is not usually marked on the carpeting itself, your dealer should be able to tell you what it is. Face weight tells you the density of the carpet, one of the most important factors in judging the quality. A dense carpet will wear better because the adjacent fibers, being closer, will support each other better. A person's weight is borne on the ends rather than along the sides of the fibers. When assessing face weight, bear in mind that in federally financed housing the minimum acceptable carpet is one with a 24-ounce face weight. More luxurious carpets have a face weight of 30 to 40 ounces or more.

The Importance of Padding   Padding adds so much to the comfort and durability of carpeting that it is inadvisable to buy carpet without also buying padding. You cannot lay conventional carpeting properly without it. Padding is often included in the cost of carpeting, but it often pays to buy a better quality. Thicker and softer padding is not necessarily better, however. Padding that's too thick or too soft can actually be harmful, since it provides insufficient support, allowing the carpet's backing to break down. Felted hair-and-jute padding is the strongest; 40-ounce padding is strong enough for the most heavily trafficked areas in the average home. For damp areas, on the other hand, foam rubber or urethane foam paddings are better.

LAYING NEW CARPETING

Either type of carpeting—conventional, which must be stretched, or cushion-backed, which may be laid down with or without adhesive—may be applied directly over almost any type of floor. It is always a good idea, however, to check the condition of the floor's understructure first, as described in chapter 2. Although carpeting will easily cover any minor irregularities, such as small indentations, be sure to set any protruding nails, fix any squeaks, and secure any loose boards (also described in chapter 2). Laying wall-to-wall carpet is easier than most people assume, since it can be pulled and stretched into place. But carpeting is expensive and the average handyperson should plan well and take his or her time. To install conventional wall-to-wall carpeting you will need to rent a *knee-kicker* and a *power stretcher,* and possibly a *seaming-tape iron.* Their use is described later. You'll also need an *upholstery hammer,* a metal straightedge, and extra utility knife blades. A *wall-trimmer* will make final cutting easier, and purchase a *row-running knife* if you have many seams or long seams to cut. When using adhesives, cross ventilate the room with open windows and doors and a window fan set to exhaust air.

Conventional Wall-to-wall Carpeting

After checking the surface of the floor, remove any shoe molding and any floor registers or escutcheon plates around any pipes through the floor. You are now ready to install the *tackless strips* and *binder bars* that will hold the edges of the carpeting securely in place.

Installing Tackless Strips and Binder Bars   Tackless strips are narrow pieces of soft plywood with rows of tacks sticking up through them, at an angle of about 60 degrees.

They are tackless since they hold the carpet neatly in place without the use of visible carpet tacks (see Figure 13). Since all carpeting is not the same thickness, be sure to install tackless strips having the appropriate size tacks—$^3/_{16}$ inch, $^7/_{32}$ inch, and $^1/_4$ inch are the three common sizes. The right length of tack can be felt through the carpet (no padding) without its actually pricking you. The strips come with nails ready to be hammered into wooden floors. You'll have to use a narrow-headed hammer—such as an upholstery hammer—so you won't bend down any of the tacks as you hammer. Nailing the strips to a concrete floor is more difficult. Wearing protective glasses or goggles, secure the strips with masonry nails, which require a heavier hammer—you'll have to be even more careful about the tack points.

Where the carpeting ends in an open space, such as in a doorway, a *binder bar* is used to hold the carpet down and keep its edge from unraveling. It is sometimes necessary to cut the ends of the binder bar to fit around the doorstop or door molding since the binder bar should be located directly under the door.

If it's impossible to nail the tackless strips and binder bars to the floor, they may be set in place with adhesive. Contact cement, used according to instructions, is usually strong enough to form a good bond between the plywood of the strips and other floor surfaces. Ventilate the room well when using it.

Start by nailing the tackless strips around the room's entire perimeter (except for the

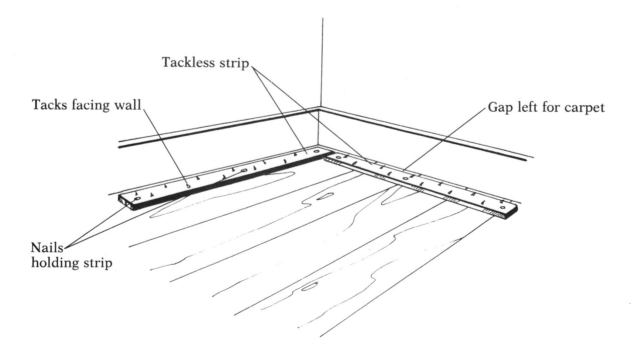

Figure 13 Tackless strips

openings, where binder bars will be used). Leave a space between the strip and the wall equal to two-thirds the thickness of the carpet to be installed. Make a spacer out of wood or cardboard to this width so that the strips can be accurately positioned. You'll be tucking the edge of the cut carpet into this gap, and it must be uniform.

Note: Nail the strips with the angled points of the protruding tacks facing the wall; otherwise they will not hold the carpet when it is stretched. The strips are soft enough that cutting them to length is easily accomplished with a pair of tin snips or heavy-duty shears— they do not have to be sawed. When you come to an obstacle such as a radiator, nail the tackless strips in front of it. After the carpet has been stretched over them, you can cut it to fall loosely around the pipes or legs.

Installing Padding When the entire area has been outlined with strips and bars, the padding is laid down within this outline (see Figure 14). It is usually made with one side designed to be uppermost; in the absence of any specific indications, the top is usually the waffled or patterned side.

Cover as much of the area as possible with each section of padding, to minimize the number of seams. Overlap the tackless strips a little; you'll trim off the excess later. Either staple the padding down (to a wood floor) or cement it down (to a masonry floor). If you cement the padding to a masonry floor, you must make sure that the masonry is dry and clean and that you use an adhesive compatible with and effective for the specific type of masonry concerned. Read the adhesive container label carefully.

After the padding has been secured to the floor, trim off the excess around the edge, using a sharp utility knife and leaving between ⅛ and ¼ inch at the edges. Any seams

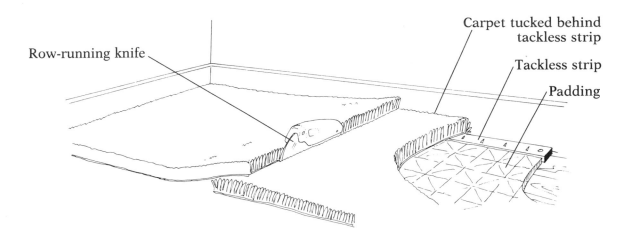

Figure 14 Trimming seams with a row-running knife

in the padding should be taped with duct tape or tape provided with the padding. Then make a final check to ensure that areas where the carpeting might slip are especially secure.

Starting the Carpet Installation   It helps if the carpet can be cut roughly to size and spread out in an adjoining room until it reaches the same temperature as its intended location. This helps to flatten it out. A flat carpet will be easier to deal with than one trying to roll up.

If there is no room to do this in the house, it may be possible to spread it outside on a dry, clean, and flat surface. Even though you are rough cutting, measure very carefully, being sure to leave about a 3-inch margin all around. If you are using two or more segments, each must be oriented correctly with regard to the pattern and pile direction. Don't make any cuts until you have marked all the measurements on the carpet. That will lessen the chance of a mistake.

Cut-pile carpet should be cut from the back. Measure along both sides of the segment, making a small cut at the proper point on each side. Roll the carpet back and snap a chalk line on the backing material from cut to cut. Use a straightedge and a sharp utility knife to complete the cut along the chalk line.

Loop-pile carpet must be cut from the face side, preferably between rows of loops. Use a firmly held straightedge and utility knife, or, ideally, a special carpet-cutting knife known as a row-running knife, along the line where the cut is needed.

Joining Seams   When the carpet is brought into the room or area where it is to be laid, the first job is to join any seams. Position the two carpet sections correctly, with one lapped over the other about 1 inch (see Figure 14). Using a row-running knife butted against the edge of the top piece as a guide, cut through the bottom piece without cutting into the padding. You will not be able to cut all the way to the walls with this knife, so complete the cut carefully with a utility knife. Remove the cut strip from the bottom piece of carpeting, then peel both pieces back to expose the padding. You can seam this joint either with *hot-melt carpet-seaming tape* or with *latex tape* and *adhesive*.

For synthetic fiber carpets, hot-melt tape is the strongest. Place a length of the tape, adhesive side up, at the point on the floor where both pieces will meet. Using the seaming-tape iron (rented from the dealer), activate the glue by slowly drawing the iron over the tape. As the tape is activated (it will take about 30 seconds at the start), lower both edges of the carpet onto the tape, making sure that the two edges butt up against each other tightly. If the pile direction of both sections is oriented the same way, the seam will be almost invisible. If necessary, keep the two edges together with weights until the adhesive has cured and the tape is bonded to the carpet.

Latex tape is similarly positioned under the seam, having been first coated with adhesive. Before lowering the two carpet sections, run a bead of adhesive along the edge of the carpet backing of both sections. Try to keep the adhesive off the fibers themselves. Press both edges firmly together. This method may be a little quicker, but the seams are sometimes more visible than those joined by hot-melt tape, and, particularly in heavily trafficked areas, they are less strong.

Stretching the Carpet   After the seams have been taped, and when you are sure the bond is secure, you can stretch the carpet and secure it to the tackless strips. Make any necessary cuts around the edges of the carpet, especially around bays and protrusions, so it will lie flat. It should be fairly accurately positioned with a safe margin (about 3 inches) all round.

Starting at one corner, and using a professional knee-kicker (rentable from tool rental stores or carpet dealers), stretch the carpet over the tackless strip. Here's how: Place the knee-kicker about 1 inch inside the tackless strip. Its teeth will grab the carpet firmly. When you nudge the end of the knee-kicker with your knee (it has a socket just for that purpose), the carpet will be pushed over the tackless strip, where it will be caught by the tacks and held securely. Angle the stretcher somewhat diagonally toward the wall. Be careful to exert just enough pressure to move the carpet—too much could damage it. Secure a foot or so of carpet along each wall running into the corner, then angle the knee-kicker away from the corner as soon as there is room to do so and secure about 3 feet of carpet away from the corner.

It's best to secure the longest edge of the carpet first, using the power stretcher (another tool you'll rent). Place the end of the stretcher against the baseboard in the secured corner, protecting the baseboard with a piece of spare carpet wrapped around a board. Engage the toothed head of the stretcher in the carpet in the corner across from where you started (see sequence in Figure 15). By pressing down on the power lever at the head of the stretcher, the carpet will now be stretched over the tackless strips in the second corner. Just as with the knee-kicker, it's best to angle the stretcher somewhat diagonally to the wall rather than facing directly into it. This helps to stretch the carpet in the direction you're going and avoids leaving bumpy, unstretched segments of carpet in already secured areas. With the power stretcher in place, use the knee-kicker to secure the carpet over the tackless strips in the second corner.

When two corners have been secured, roll the edge of the carpet between them over the tackless strips. Then, using the power stretcher, secure the corner diagonally opposite the corner you secured last. Now you have a third corner secured and can roll the edge of the carpet over the tackless strips on a second side. Three corners and two sides will now be secure—as in Figure 15.

Finally, place the back of the power stretcher again in the first corner and work your way down the length of the room, going alternately from one wall to the other so that the fourth corner will be secured. You may want to use the knee-kicker to roll the carpet over the tackless strips in this last corner.

Finishing the Job   When the carpet is all stretched flat and secured, you only have to trim the edges. Do this very carefully with the specially designed carpet wall-trimmer. The

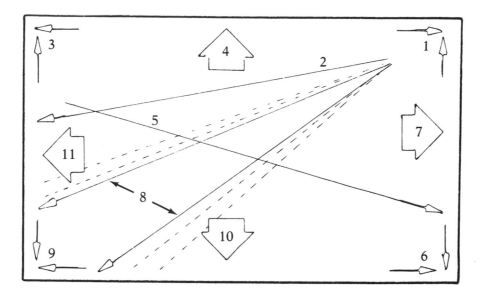

1, 3, 6, 9: Use knee-kicker to stretch carpet over tackless strips
2, 5, 8: Use power stretcher to stretch carpet into opposite corners
4, 7, 10, 11: Roll edge of carpet over tackless strips between secured corners

Figure 15   Order of carpet-stretching operations

trimmer, which can be adjusted to the exact thickness of the carpet, makes it easy to cut along the edge of the carpet leaving just enough to be stuffed down into the gap between the tackless strips and the walls. The easiest tool with which to wedge the edge of the carpet down is a broad-bladed putty knife, but for narrow sections and especially stiff carpet you may find a fairly large screwdriver better. You'll still have to use a utility knife in corners and around obstacles such as radiator pipes. At a binder bar (if you've used them), trim the carpet off at the middle of the bar. Hammer the binder down over the edge of the carpet, using a block of wood between your hammer and the bar.

The job should now be complete. In certain cases, it may look better to install (or reinstall) shoe molding over the carpet around the baseboard. In most cases, however, it's better to omit the molding.

Cushion-backed Carpeting

The great advantage of cushion-backed carpeting is that it does not have to be stretched when laid. Furthermore, since the padding is already bonded to the carpet, there is no need to buy or install a separate layer of padding. But despite the saving in labor and original cost—it is invariably cheaper than conventional carpet—its use is not always appropriate.

You can install cushion-backed carpeting either with adhesive or with two-sided tape.

Both require that the underlying surface be perfectly clean to obtain the best adhesion. Using adhesive produces the best results, but if the carpet ever needs to be taken up, the backing will be ripped off the carpet, rendering both carpet and floor unfit for further use.

Installing with Adhesive   Before applying any adhesive, cut the carpet to size, allowing at least 2 inches extra around the edges and at every seam. This is best done using a scaled drawing of the area to be carpeted, spreading the carpeting out on a clean, dry, flat area, and transferring all measurements to the carpeting before making any cuts (see page 36). Be sure that adjacent sections match in pattern and in pile direction. Make all cuts from the top side of the carpet, using a sharp utility knife and a straightedge.

Bring the carpet or sections of carpet to the area where they are to be installed. Join any seams first. Using the carpet sections as a guide and the walls as a reference point, measure carefully and snap a chalk line to mark the seam location. Carefully position the edge of one section of carpet on the marked line, then overlap it with the other section approximately ¼ inch. Roll back both sections and, with a *notched trowel* made especially for spreading adhesive, spread adhesive so that it covers the floor for about 1 foot on either side of the marked line. Don't obscure your chalk line.

Now, replace one carpet section so its edge is aligned on the marked line. Apply seam adhesive along the seam edge of both sections' backing, being very careful not to get adhesive on the pile. Butt the second section of carpeting up against the first and press both firmly into place, working away from the seam so that all air bubbles are squeezed out beyond the area covered with adhesive. Allow this seam area to dry thoroughly (see label instructions on the adhesive).

When the seam area is dry, roll the carpet sections up from the ends opposite the seam, to expose the surface of the floor. Apply more adhesive to the floor, working from the area next to the seam out toward the edges of the area to be carpeted. Unroll the carpet as you go, smoothing it down and working out any trapped air bubbles as before.

When the whole area has been covered, the outer edge should be trimmed, with an overlap of no more than one carpet thickness. This extra should then be tucked down against the wall using a broad putty knife. Edges of the carpet that do not abut any walls should be finished with binder bars, as described in the previous section on laying conventional carpeting.

Installing with Double-faced Tape   The use of double-faced adhesive tape is easier but less secure, especially in areas of heavy traffic. The carpet may shift and seams may open. However, a tape installation may be perfectly satisfactory where traffic is light or permanence is not required.

The tape installation starts the same way as an adhesive installation: Cut the carpet to size, but before bringing it to the area where it is to be installed, apply a band of double-faced tape around the entire perimeter of the area, leaving the second protective layer still on the tape. Fit the carpet into place but allow it to lie there for at least a day before sticking it to the tape. That will allow the carpet to relax and flatten, putting less strain on its adhesion to the tape. Remove the protective backing from the tape and press the carpet down onto it. After trimming the edges, any that are exposed may be bound with carpet-edging tape or secured under binder bars, as described on page 58.

CARPET MAINTENANCE

Regular Cleaning

Deterioration of carpeting results not so much from simple foot traffic as from the action of ground-in dirt and grit abrading and wearing the fibers. The best thing you can do to preserve your carpet, therefore, is to keep it clean. Loose carpets and rugs can be taken outside and beaten, to dislodge and shake out all the small dirt particles that cause most of the wear. With wall-to-wall carpeting, a vacuum cleaner in good condition and with sufficient power can do a satisfactory job. But it must be used regularly. In an average household, two cleanings a week, with perhaps localized cleanings of heavily used traffic lanes, should suffice.

Heavy-duty Cleaning   Even with regular vacuuming, it is a good idea to treat the carpet to a more thorough, overall cleaning once in a while—perhaps at yearly intervals—to restore freshness and color. There are three ways this can be done: with wet shampoos; with hot-water-extraction methods; or with a dry-cleaning agent.

Shampooing is the easiest method: An electric rotary brush works the detergent into the carpet pile, loosening the dirt and suspending it in the foam. When the solution has dried, the residue containing the dirt is vacuumed away. You can rent all the equipment necessary at many places. The job is not hard to do, but there are some precautions: You must be careful not to abrade the pile by excessive or too vigorous an application, and you must not allow the carpet to become too wet or the padding or backing may suffer.

The *hot-water-extraction* system (often referred to as steam cleaning) is better. It should be used instead of wet shampooing, if not at every cleaning, then at least every third time. With this system there is no mechanical brushing to damage the pile. The machine dispenses a nonfoaming detergent solution into the carpet under pressure, then immediately sucks most of it back again together with the loosened dirt. You still have to guard against the possibility that the carpet may become too wet, with resulting damage to the fibers and the backing.

Both shampooing and the hot-water-extraction system usually require that the carpet be allowed to dry for at least two or three hours after it has been cleaned. That is not necessary if *dry cleaning* is used. In this method, you brush a special powdered compound down into the pile. That loosens the dirt; then you vacuum thoroughly, bringing up compound, dirt, and all. This method is not considered as effective as either of the wet methods.

Emergency Spot and Stain Removal

The number of chemicals in use around the home today—from acne prescriptions to preservatives in soft drinks—combined with the various fibers in carpeting make the prevention and elimination of stains extremely difficult.

The next best thing to prevention is immediate action. If left unattended, certain substances can produce irreversible damage. A comprehensive list of possible staining agents and their treatments is impossible, but here are a few general hints:

- When diluted properly, dishwashing detergent will remove most water-soluble stains.
- Acid stains—such as soft drinks, fruit juices, and vomit—may be neutralized by a solution of ammonia (1 tablespoon of clear household ammonia per ½ cup of water), but do not use this on wool carpeting, as it will only help to set the stain.
- One part vinegar mixed with two parts water helps deal with smelly stains.
- Carpet shampoo is also useful to have on hand and can be tried on any stain.

If any area becomes excessively soaked, the excess should be blotted up with paper or other towels rather than allowed to evaporate.

Spot-removing fluid works well on grease and oil stains, but use it only with great care. Many spot-removing fluids contain perchloroethylene, methylene chloride, or both, and these substances are known to cause cancer in animals. In addition, their vapors are harmful to inhale. If you use a spot-removing fluid, cross ventilate the room well: Open all doors and windows and use a window fan set to exhaust air. Avoid inhaling the vapors. Use the fluid only when necessary, use as little as possible, and keep the cap closed as much as possible; open the bottle only to apply the fluid and close it immediately. Of course, all household chemicals should be kept where children cannot reach them. For particularly difficult problems the only solution may be to have the carpet professionally cleaned.

6

CONCRETE FLOORS

ASSESSING THE CONDITION

Many houses are constructed right on concrete slabs, many finish floors are laid over concrete, and many floor areas in and around the house remain concrete, sometimes painted, sometimes textured, and sometimes unadorned. In short, concrete is a much-used material; there is almost certain to be a concrete floor in your living space.

In itself, concrete is a remarkably stable and long-lasting material; problems that develop are usually the result of exterior factors rather than of any intrinsic fault. When sizing up the condition of any concrete floor, therefore, be aware that problems like cracks, movement, dampness, and peeling paint probably have more to do with underlying causes than with the concrete itself.

You may occasionally encounter concrete that has been poorly mixed, made with poor quality or the wrong ingredients, or made with the wrong proportions of ingredients. The concrete may be crumbling, powdering excessively, or otherwise falling apart. There is little that can be done for concrete in this condition; the only remedy is removal and replacement. Short of such a problem, however, are many less serious conditions you *can* deal with—cracks, gaps, and small holes.

Catastrophic damage, such as that caused by a landslide or earthquake, either by sudden shaking or as a result of slow but continual creeping, can also compromise the integrity of concrete structures if they have not been adequately reinforced, even though the material itself remains in good condition. Any structure damaged in this way will probably have to be replaced, since it is very difficult to attach new concrete to old.

REPAIRS

Before trying to repair any cracks, gaps, or shattered areas of more than 1 foot in length, try to determine the basic cause. If necessary, consult a qualified building engineer. Unless you know there is no underlying problem, your repair will be short-lived. You may also be ignoring a more serious problem that should be addressed, such as subsidence, creep, or even imminent collapse—the result of such things as insufficient ground preparation, poor drainage provisions, or inadequate reinforcement within the slab itself.

Cracks of some length in foundation slabs may be the result of settling caused by the weight of the house on an improperly formed foundation. That may have occurred long ago and long since stabilized. On the other hand, it could be evidence of current conditions that need attention, such as poor drainage or ongoing deterioration of other foundation elements. Small cracks may simply be the result of small irregularities in the overall mix of a slab—a cosmetic problem.

Large cracks in patio areas and outside walkways are invariably the result of inadequate preparation of the poured slab. If these areas are not part of the structural framework of the house, you can certainly repair them without dealing with the underlying causes. It may not be worthwhile to tear up a whole walkway just to level it perfectly, when filling a few cracks would be adequate.

Small cracks in outdoor areas are more probably the result of repeated cycles of moisture freezing and thawing. If left unattended, however, these can turn into serious cracks over time. Each time a crack is formed, more space is created for moisture or water to collect and freeze, thereby creating an even bigger crack.

Minor concrete repairs require little skill. Any willing handyperson can undertake the task and get good results. For special tools, you'll need safety goggles, a soft-headed steel hammer, a square trowel, and a large bucket, at least, in which to mix patching compound.

Filling Cracks

Concrete is made of sand, portland cement, and gravel in varying proportions according to varying needs, mixed together with water and allowed to cure (set and dry) over an extended period. Once cured, the addition of new material is extremely difficult. For one thing, the water used to create the added concrete tends to be absorbed by the old concrete before the new concrete has had time to cure properly. Furthermore, the coarse gravel in the new material prevents a good bond between the old and new concrete. These problems have made concrete repair very difficult until very recently. Now, new compounds containing epoxy and latex, among other things, have alleviated that situation.

Preparing the Crack   Any crack you want to repair must first be thoroughly cleaned of all loose material, dirt, and debris. If the crack is large enough to take the end of a cold chisel, you should *undercut* the edges of the crack (make it wider at the bottom than at the surface) so that the repair will hold much better (see Figure 16). Wear safety goggles, and use a soft-headed steel hammer. To ensure the best possible bond, you should thoroughly moisten the area and keep it moist for several hours before filling it.

Commercial Patching Materials   For medium cracks or holes there are three types of concrete repair material readily available at hardware stores:

- *Latex cement,* which is supplied in two parts, the cement and a latex binder, and mixed together as needed to form a thick paste that can be troweled into the crack and smoothed off.
- *Vinyl cement,* which requires just the addition of water, is then used the same way.
- *Epoxy cement,* which comes in three parts: dry cement, an emulsion, and a hardener. The emulsion and the hardener are first mixed together, then the cement is stirred in until the mixture has a pastelike consistency.

With all three patching materials, it is important to adhere to the drying instructions on the packaging, since this will determine how long the repair lasts.

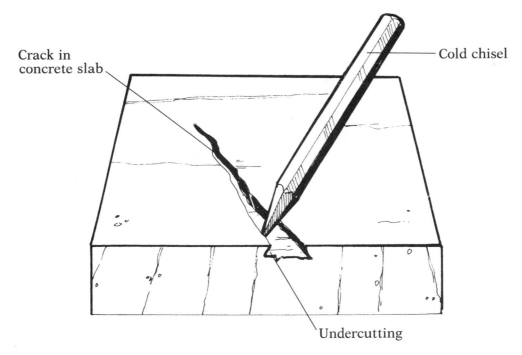

Figure 16   Repairing a small crack

Mixing Your Own Patching   You can make your own patching material, although commercial preparations are more convenient. For the smallest holes and cracks, mix portland cement and water so that the paste will just hold its shape. A similar material is cement-base ready-mix grout, usually sold in cartridges designed to be dispensed with a caulking gun. Using a caulking gun premix is ideal for small cracks; just squirt in the material and smooth it over with a trowel.

For larger cracks, make a heavier mortar mix with one part portland cement, three parts sand, and enough water to produce a well-mixed but firm consistency. To help bond the new material to the old, paint the edges of the crack with a paste made of portland cement and water. Do not allow the paste to dry before applying the mortar mix; have both ready to use before you begin the repair. Trowel the mortar into the crack, smooth it off, then let it dry until you can just leave a thumb print. At this point, cover it with plastic (weighted down at the edges), and leave it for the better part of a week, removing the plastic daily to dampen the patch. In this time, the patch will cure properly before drying out.

Repairing Flaking Concrete

One of the advantages of epoxy mixes—which are also good for filling larger cracks—is that they do not require such an extended curing time, usually becoming completely hard within twenty-four hours. They are also excellent for repairing surfaces that suffer from *flaking*, which can happen to concrete that has been troweled to a smooth finish. The smooth surface results from a top layer of concrete made without stones and gravel (called *aggregate*). While this has a more finished appearance than untroweled concrete, it is more susceptible to frost and water damage. Moisture, seeping below the surface, causes the smooth layer of mortar to flake away from the aggregate concrete below.

To restore a flaking surface, you have to remove all the flaking and scaling, breaking it up further if necessary with a ballpeen hammer (wear protective goggles to guard against flying chips and fragments). If the area is large, you may have to use a sledgehammer. Remove all loose debris and remaining particles with a wire brush and a broom. Then soak the area, keeping it damp for several hours, before applying the epoxy mix, prepared according to the package instructions. A square trowel, rather than the usual pointed trowel, is best to use when filling and smoothing. Work from the center out, feathering the edges (smoothing the new patch material into the old concrete around the patch). When you're satisfied that the surface is smooth and level, no further preparation or curing is necessary. Simply wait for twenty-four hours before walking on it.

Patching Larger Areas

Larger damaged areas can also be repaired, but this usually involves replacement rather than simple filling. The work can be laborious, but requires minimal skills. Complete removal of the concrete in the damaged area is usually the hardest part of the job. Break up the damaged area with a sledgehammer (wear protective goggles and a stout pair of gloves). Since you'll have to break the concrete into lumps small enough to be carried away, it may even be worthwhile to rent an electric jackhammer. They're made in all sizes and, while noisy, are efficient. A jackhammer can do in seconds what a sledgehammer might take half an hour to accomplish.

Preparing the Hole   When all loose concrete has been removed, use a hammer and a cold chisel to cut the sides of the remaining concrete so they slope inward toward the center of the hole (see Figure 17). Now dig out the earth at the bottom of the hole a few inches deeper than the surrounding concrete. Tamp this area firmly (with the end of the sledge-hammer or a two-by-four), then fill the hole with gravel or split rock (purchased from a building supply center or lumberyard) up to the level of the bottom of the surrounding concrete. The gravel bed provides drainage and will help prevent the new patch from heaving, keeping it level with the surrounding material.

The next step—reinforcement—is very important, and its omission may well have been the cause of the damage in the first place. You must suspend some form of reinforcement in the hole, ideally about midway in its depth and reaching completely to the entire perimeter. Reinforcement can be in the form of different-sized rods (called *rebar*) or mesh (known as *reinforcing mesh*). For most purposes, mesh is easier to use. It is available in rolls 5 feet wide. With metal shears, cut out enough mesh to cover the area needed, joining adjacent separate sections, if necessary, by twisting the ends of the mesh together or by securing them with wire ties. Use stones or bricks or small pieces of the broken-up concrete to support the mesh halfway up the hole.

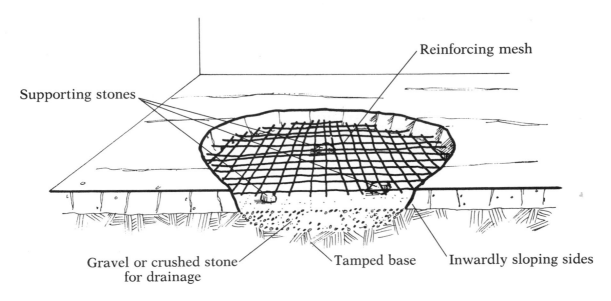

Figure 17   **Large patch preparation**

You are now ready to pour your own minislab! If the area is relatively small, it may be most convenient to use a bag of ready-mix concrete, obtainable in 25-pound bags from hardware stores or lumberyards. If the repair is large, it may be easier to mix your own, since separate ingredients are considerably cheaper.

Making Concrete Concrete consists of gravel or small stone (called the aggregate) suspended in sand rougher than the fine, uniform sand used for mortar. The sand for concrete consists of grains that vary in size from very small to about $1/16$ inch in diameter, the better to fill in the irregular spaces between the gravel and stone of the aggregate. The aggregate and the sand are bound together by portland cement mixed with water. When all these ingredients are properly mixed in the right proportions, an almost indestructible substance called concrete is formed.

The aggregate is sold by the cubic foot. If there is a choice of size, choose the smaller size stone (up to 1 inch in diameter) for small repairs, and the larger size stone (up to $1\frac{1}{2}$ inch in diameter) for more extensive repairs, such as to walks and patios more than 4 inches thick. The aggregate should be clean.

Sand is also commonly sold by the cubic foot. Choose the rougher type, as mentioned, for concrete. Do not use beach sand, even though it may be yours to take; the salt content will compromise the concrete's strength.

There are many different types of cement available, but you need only be concerned with two—plain portland cement and type 1A portland cement. They are available in bags that contain 1 cubic foot and weigh 94 pounds. Type 1A cement must be used for anything more substantial than small repairs or post anchors. It contains additives that produce and trap bubbles of air in the concrete. The bubbles provide space for the concrete to expand and contract, which helps deter cracking and makes it more weather resistant. Many building codes require the use of this type for all outdoor concrete.

When buying bags of cement, make sure they are not torn and that they contain no lumps. Cement can lose its freshness, and if it has absorbed too much moisture, the powder will have hardened. Hard areas around the edge of the bag may be crushed into powder, but you don't want cement that is hard to mix because of excessive exposure to moisture.

Water is the simplest ingredient, but too much or too little can render the mix useless even if the other ingredients have been mixed in the correct proportions. Basically, the proportions are as follows: one part cement, two parts sand, and four parts aggregate, mixed together with enough water to produce a mix that might be described either as just fluid enough to pour into a form or as firm enough to be molded without excessive slumping.

Recipes for different applications are invariably printed on the cement packaging, but these assume the use of so-called wet sand—sand that is moist enough to be balled up without exuding any moisture in your hand. Since sand is seldom in exactly this condition, the amount of water you add must be varied commensurately.

You may want to rent a mortar box and masonry hoe. Small amounts of concrete can also be conveniently mixed in a wheelbarrow or on a large board, such as a piece of plywood.

If you need more than can be made in one or two batches in a wheelbarrow, it will pay to rent a gasoline or electric mixer. A power mixer not only makes lighter work of the job, but it ensures a thorough mix, which is very important. If you are working on a very large job—such as pouring a complete walkway or a basement floor—you should consider ordering ready-mixed concrete delivered by truck.

The usual procedure is to measure out the correct proportions of the dry ingredients—sand, cement, and aggregate—mixing them into a small hill. Make a hole in the center, then add the water, bit by bit, mixing thoroughly with a long-handled spade or hoe until the correct consistency is reached.

Pouring the Concrete   Before shoveling the freshly made concrete into the repair, coat the sides of the existing concrete with a concrete bonding agent. You can use a cement paste made from portland cement and water, but there are also commercial preparations that may be more convenient. Begin filling the hole immediately, taking care that the concrete is spaded well into all corners and completely surrounds the reinforcing mesh. Make certain you leave no large air pockets. The hole should be filled to the level of the surrounding slab, with an extra shovel or so added to compensate if there is any settling in the first few minutes.

To level the new concrete, use a straight two-by-four long enough to reach completely across the patch (see Figure 18). Move it across the whole area, jiggling it up and down as you go. Make sure that any depressions not touched by the two-by-four are filled, then relevel the whole surface. When the patch is thus smoothed out, a thin film of water will usually form on the surface. As soon as this begins to evaporate you will be able to trowel the whole area smooth. Use a square trowel, not a triangular trowel. If you want a textured surface (for better traction), use a stiff broom to sweep the surface, but only *after* the concrete has become somewhat firm.

As soon as the concrete has hardened, usually within an hour, keep it covered and damp for a week. The easiest way to do this is with a sheet of plastic—lifting it off to sprinkle more water on the concrete whenever necessary to keep it damp.

If you've overestimated the amount of concrete needed (and it's always better to overestimate than underestimate), the best way to dispose of the excess is to separate it into lumps small enough to be carried away when dry. If you have a lot left, you can pour it into forms to make blocks you can use for walls or stepping stones.

PAINTING CONCRETE FLOORS

Raw concrete, especially if the finish has been textured, can have a certain rough-hewn appeal in outdoor situations, but it usually presents problems if left untreated indoors. It creates dust, it is unduly susceptible to spills and stains, and it may be the wrong color or simply look too unfinished.

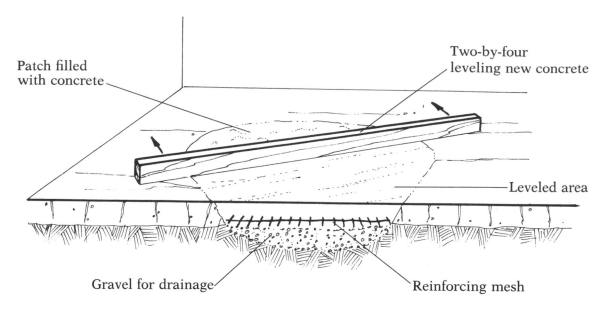

Figure 18   Leveling a concrete patch

If you aren't planning to install a finished floor (wood, tile, or carpeting) over the concrete, there are two ways to deal with it. One is to treat the floor with a *penetrating sealer,* which does not change the color but imparts a certain luster. Sealer also eliminates the dust and provides some protection against spills and stains. The other is to paint the floor with a special *concrete paint.* Neither sealing nor painting a concrete job is particularly difficult. But take care to cross ventilate the area you're working in, which is sometimes difficult in a basement. Open as many windows and doors as you can, and use a window fan or two set to exhaust air. Plan ahead: Don't paint yourself into a corner.

Preparing the Concrete

Prepare concrete for painting or sealing by a thorough cleaning. Neither paint nor sealer will adhere well to dirty surfaces, and a sealer will not conceal oil and grease stains. Start by sweeping away and vacuuming up all debris, dust, and loose pieces. If the concrete has been float-finished—troweled to a very smooth finish—this top surface may not adhere very well to the concrete below. You may have to wire-brush the surface until no more dust or powder comes off.

Oil stains and grease spills must be scrubbed thoroughly with a strong detergent. If that doesn't work, use trisodium phosphate (available at hardware stores), dissolved in water. Trisodium phosphate is an irritant: Always wear safety goggles, long sleeves, and rubber gloves when working with it. Iron stains should be treated with a mixture of one part oxalic acid to nine parts water. Again, take precautions when using oxalic acid: Wear goggles, long sleeves, and rubber gloves. Copper or bronze stains should be removed with a solution of one part ammonia to nine parts water. Other persistent stains may sometimes be removed by scrubbing with mineral spirits (paint thinner). Cross ventilate the area when working with acids, ammonia, or solvents.

Previously painted areas can be repainted if there are no areas of peeling paint. Glossy surfaces must be lightly sanded so the new paint will adhere better.

Taking Care of Dampness   One of paint's worst enemies is dampness—the paint simply won't stick for very long. Rooms with concrete floors may sometimes feel damp because of condensation. Sufficient ventilation and a little heat in the room should correct this condition. But if the dampness is a result of underlying drainage problems—moisture may be seeping up through the concrete—the solution will probably call for professional help to waterproof the outside of the building. It may be necessary to deal with more fundamental drainage problems.

A simple test for moisture in a concrete slab or floor is to tape a square of plastic to the floor, sealing every edge. Leave the plastic in place for twenty-four hours, then pull it up and examine it. If there is moisture in the floor, condensation droplets will have appeared on the underside of the plastic.

Efflorescence, a white discoloration or powdering caused by salts in the concrete leaching out, is an ominous sign of inherent dampness. It will almost certainly require professional treatment.

Sealing and Painting   Since concrete sealer actually penetrates the concrete to a certain depth, there is little concern about the adhesion of the sealer to the concrete surface. Sealers are relatively inexpensive, and application is straightforward. Sealed concrete will be dust-free and adequately protected from stains, assuming spills are promptly wiped up.

The Alkali Problem   Painting is trickier since the paint does not penetrate the concrete and must adhere firmly to the surface. That's why it's so important to start with as sound and as clean a surface as possible. That's true of any material to be painted, of course, but concrete offers some special problems.

Concrete contains alkalies that can adversely affect paint, causing it to peel and blister. To neutralize these potentially damaging alkalies, many paint manufacturers used to recommend washing the surface first with muriatic acid. This also etches the surface and increases paint adhesion. Muriatic acid is readily available and often used for masonry cleaning and restoration. But it is extremely hazardous: Inhaled vapors are harmful, skin contact can result in burns, serious eye damage is possible, and plant life can be killed if it seeps into the garden. Avoid using it. There are now concrete paints designed for application without the prior use of muriatic acid.

Different paints do different jobs, so assess your situation and read the labels carefully before purchasing a particular type of concrete floor paint.

Choosing and Applying the Paint

The toughest paints and those with the best adhesion are the *epoxy-based polyurethane paints*. Be sure to choose the so-called two-part epoxies—they come in two separate cans and call for the resin and the hardener to be mixed before use. The one-container "epoxies" may not be true epoxies and certainly do not work as well. Epoxy-polyurethane paints can be applied as soon as a month after concrete has been poured, usually require two coats, and dry within 6 hours to a hard, glassy finish. Once mixed, two-part epoxies give you only a limited time in which to work. They are also the most expensive. Unfortunately, they can be slippery when wet, and noxious when applied—make sure you have adequate cross ventilation.

A specialized paint for those areas that have potential dampness problems is *rubber-based paint,* which is solvent-thinned. It resists moisture well, although it does not last as long as epoxy paints. Furthermore, application is more complicated, requiring three coats, applied at long intervals. The solvents involved are also toxic: These paints should only be used if you wear safety goggles, a respirator, and rubber gloves and you create adequate cross ventilation.

Water-based latex paint, although not as long lasting as epoxy or rubber-based, is much cheaper, much easier to apply, and not as slippery (because it's not as glossy). Latex usually requires three coats for adequate coverage and should not be applied until at least three months after the concrete has been poured. It does not withstand heavy traffic as well as epoxies and rubber-based paints and is susceptible to damage by freezing, which makes it better suited for lightly traveled indoor use.

There are also innumerable *oil-based paints* available in a great range of colors and at considerably less cost. But they don't generally last as long as the more expensive paints, primarily because of their poor resistance to the alkalies in concrete. They will last longer if you use a muriatic-acid wash first, but you *must* take into account the dangers involved, as described above. An alternative, which is somewhat less effective but far safer, is to apply a concrete sealer as an undercoat.

One choice for areas that are likely to see rough and dirty wear is spatter paint, since the pattern hides marks and stains much better than plain paint. To apply any of the paints, use a roller on a long handle. If the concrete surface is rough, use a shaggy roller especially made for rough surfaces.

7

STAIRWAYS

BASIC STAIRWAY CONSTRUCTION

Although *stairway* usually refers to the place where stairs are located, we'll use it to refer to everything connected with stairs: the stairway, the actual stairs, and any landings.

Stairs made of masonry or metal require little maintenance; and should they become damaged, there is little that the average do-it-yourselfer can do. Certain problems of wood stairways, however, can be successfully repaired by the amateur, providing the basic construction is understood.

By the end of the last century, stair-building had evolved into a complicated and specialized art. The geometry involved in planning curved and sloping woodwork, and the skill required for its execution, required artisans of great ability with considerable training. Although few of today's houses feature sweeping stairways with carved handrails and spandrels, and ornately turned balusters (often called banisters), newel posts, and balustrades, the basic construction of most stairways today derives directly from this period. As a result, for all but the simplest of houses, stairways are usually manufactured and supplied by specialist firms following (in simplified form) the principles developed in the nineteenth century.

The Parts of a Stairway

All stairways share certain components in common. The piece you stand on is known, logically enough, as the *tread*. The vertical piece that rises between one tread and the next, forming the back of the step, is known as the *riser*. Treads and risers (together forming the actual steps) are supported by the *carriage*. This is a system of two or more timbers running the whole length of any particular flight of stairs, each member having its own name, such as *wall stringer* or *carriage string board*. Unless a flight of stairs is located between two walls, any open side will be enclosed by a *balustrade* consisting of a *handrail,* supported by individual *balusters* (or banisters), terminating in a *newel post* at one end at least.

Other parts, which may or may not be included in a particular stairway, are *landings* (small floor spaces connecting flights of stairs between the floors of the house), *winders* (wedge-shaped treads found where the staircase turns corners), and various other, mainly ornamental, elements such as *nosings, moldings,* and *return moldings* (see Figure 19).

Different Types of Construction Since stairway construction can range from the very simple to the exceedingly complex, it is hard to define even broad categories, but some distinctions will make certain repairs easier to understand and undertake. The first is between *open* and *closed* stairways. The difference is whether the back of the treads and risers are visible and accessible (an open stairway) or not (a closed stairway). Just because the area under the stairs is enclosed (by an under-the-stairs cupboard, for example) doesn't necessarily mean you won't be able to get at the treads and risers from behind. On the other hand, a freestanding stairway may well be paneled or otherwise finished behind, preventing direct access to the back.

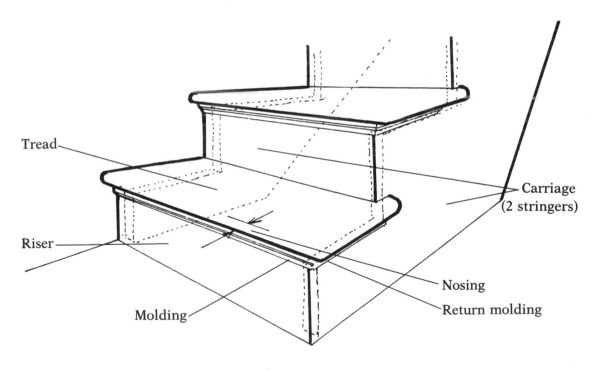

Figure 19 Staircase parts

The next important distinction is between the methods used to attach the treads and risers to each other and to the carriage. In the simplest method, the treads rest on step-shaped stringers, with their back edges butted against the risers, which in turn are butted up under the treads above them. The various parts are held together either by nails or screws. More sophisticated methods involve joining the treads and the risers through various forms of housed joints, or inserting the ends of the treads and risers into joints cut in the sides of the stringers. The latter method invariably includes the use of wedges as well.

ASSESSING THE CONDITION

Before considering the possible repairs you can make, you have to take a look at the stairway system as a whole.

Most stairways were originally built true and sturdy. If a stairway is rickety or no longer true, it doesn't necessarily mean the stairway was poorly built. Settling of the house, subflooring that has rotted, or weak bearing or retaining walls may have taken their toll on the stairway. If this damage has since been rectified, or at least stabilized, then there are things you can do to improve the bearing and condition of the stairway.

A Tilted Stairway

For example, a severely tilted stair may be caused by sagging of the supporting joists. Newel posts, to which both stringers and balustrades are attached, frequently go through the floor on which the staircase rests and are attached to joists below. If these joists have sagged, so will the stairway. Faced with a leaning stairway, then, first inspect the floor substructure, as described in chapter 2. It may be possible to straighten the stairway, for example, by jacking up and reinforcing the subfloor.

The problem may lie with the walls to which the stairway is attached. Inspecting these will reveal whether the sagging is the result of an intrinsically weak stairway or of a fault in the surrounding framework.

Next, check the stair carriage. A weakened stringer can be reinforced and a rotted end can be replaced, but if the whole affair is compromised by damage or decay, it is probably better to have a new stairway built.

Treads and Balusters

One or two treads can be replaced with varying degrees of difficulty, but if most of them are in serious need of repair, it's probably best to start over with a new stairway. Keep in mind that very old stairs may well have worn and even bowed or buckled treads without indicating a weakened or dangerous condition. If the carriage and the actual wood seem sound, it's probably best to treat this simply as part of the character of an older house.

The balustrade is also subject to certain damage without reflecting on the basic condition of the rest of the stairway. One or two balusters and handrails can be replaced, and even loose or damaged newels can be repaired. Before starting on any of these repairs, however, it would be wise to assess the condition of the stairway as a whole. By and large, repairing stairways requires skill and patience. Anyone undertaking repairs should be competent and experienced in working with wood and using woodworking tools.

REPAIRS

Squeaking Stairs

A squeaky staircase, if it's open construction (with a visible and accessible back), is one of the easier problems to fix. Not only will you be ridding yourself of an annoyance, but the chances are that in fixing the squeak, you'll be strengthening the stairway (see Figure 20).

Position yourself under or behind the stairway and watch what moves when someone walks on the part that squeaks. You'll probably be able to see a tread or a riser bend—that's your squeak! Wood is continually shrinking and expanding due to changes in the moisture content of the air. Since a stairway is made up of many separate pieces of wood, often with their grain oriented differently, it is inevitable that sooner or later some parts

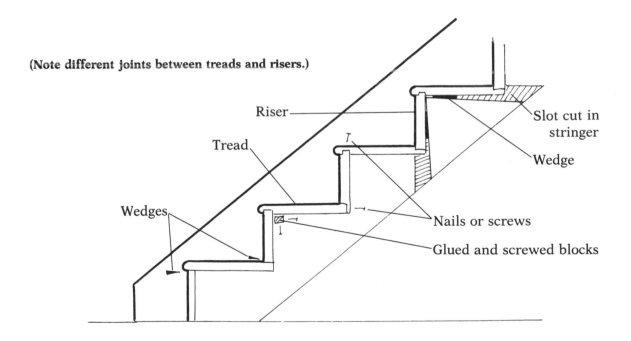

(Note different joints between treads and risers.)

Riser

Tread

Wedges

Slot cut in stringer

Wedge

Nails or screws

Glued and screwed blocks

Figure 20   Squeak prevention

will shrink more than others. Gaps form, and when these are closed by pressure on the stair, the separate parts rub against each other, thus producing squeaks.

Lubrication   An immediate—but usually temporary—remedy is to lubricate the two pieces that rub, using powdered graphite, talcum powder, or even floor or finish oil. Wipe up any excess before it has a chance to stain anything. This can easily be repeated, but if it doesn't last long, it is best to try a more permanent repair.

Using Wedges   Attempting to glue two moving parts together is not likely to succeed— the forces that caused the wood to shrink are more than strong enough to break any glue joint. You simply risk making a mess.

A better way is to stop the movement of the two pieces. If the stairway is made with treads and risers fitted into slots in the stringers, they will usually be held in these slots by wedges (Figure 20). Tap the wedges in a little further until you've prevented any movement.

If the stairway doesn't seem to have been built using wedges, you may still be able to insert your own at the appropriate places. But be careful not to drive them in so far that all you accomplish is to push the two pieces farther apart—you only need to prevent movement.

Using Glue Blocks   Whether you'll be able to insert wedges at the critical spots will depend on how the treads and risers are attached. If they are simply butted together, a wedge can be inserted, and any protruding excess cut off. But if they are otherwise jointed, wedges may not work. In this case, attach two-by-two blocks of softwood, about 4 to 6 inches long, in the right angle formed by the two moving parts (Figure 20). Glue blocks in place and secure them with screws or nails. *Note:* You must make pilot holes, since treads and risers are usually made out of hardwood that would be difficult to nail or screw into directly. And be careful not to use nails or screws that are too long!

Use of blocks is only possible if the squeak results from a tread moving up and down on the riser below. If the squeak is caused by the bottom of a riser moving against the back edge of a tread, the cure is to use nails or screws to secure the bottom of the riser to the end of the tread.

Closed Construction   If you can't gain access to the back of a stairway, you may still be able to prevent movement by nailing through the surface. Use two nails sloping toward each other (Figure 20). These should, of course, be predrilled and then filled.

Still another way of dealing with squeaks, when the construction is closed, is to provide support to the moving parts by adding extra molding nailed under the front edge of the treads or in the corner formed by the bottom of the riser and the back of the tread. For aesthetic reasons, if you add molding at one tread, you should add it at all the others.

Replacing a Tread

Occasionally, one tread will be damaged to a hazardous point, even though the rest of the stairway is in satisfactory condition. Replacing the tread can be difficult, but it's worth doing as opposed to installing a whole new stairway. Simply installing a new tread over the damaged tread is just as dangerous as leaving a hazardous tread unattended to. The difference in the height between steps can cause the unwary to stumble dangerously, and could precipitate a fall.

If the following description of the replacement procedure is not applicable to the stairway in question, or if you don't think you can carry out the repair satisfactorily, call in a professional. Don't live with a defective stairway. You might accommodate to the danger, but sooner or later it will trap someone else.

Removing the Old Tread　If the stairway consists simply of two stringers and open-backed treads, with no risers, the repair is easy. (This type of stairway is the most vulnerable to tread damage since risers offer a lot of support to the whole structure.) First, remove the tread in question by sawing through the middle of it from front to back and removing the pieces. This may involve removing all nails, screws, or bolts. Locate these by inspecting both sides of the tread, top and bottom, and the stringers on each side. It's possible that the tread may have been nailed or screwed from the outside of the stringers, with the nail heads or screw heads set and plugged. You'll have to unplug them to remove them. If the tread has been fitted into slots cut on the insides of the stringers, and further secured with wedges, remove and save the wedges.

Getting a Replacement　If the tread was fitted into slots, measure the distance between the inside faces of the slots, *not* the distance between the inside faces of the stringers. If it's possible to insert a new tread into these slots from either the back or the front, get a replacement tread that is exactly this width. If it doesn't seem possible to do that, get a tread that is as wide as the distance between the stringers plus the depth of one of the slots.

Take your measurements (and the remnants of the old tread) to the nearest lumberyard. If they don't have an exact replacement in stock, or can't make one for you, they can probably recommend a nearby stair builder who can. Try to find a replacement that matches the old one not only in width and depth but also in thickness and type of wood. Hardwoods, such as oak, are the most common material used; it's important to get a replacement with the same strength, even if you don't have to match the grain and color, for example, because you're painting or carpeting the steps.

Installing the Replacement　If a full-width replacement can be slid into the original tread's grooves in the stringers, do so, fastening the replacement as the original was. Make sure that the tread is held firmly in place; if in doubt, use more nails or screws. Drill pilot holes.

In some cases, the replacement cannot be slid into its slots and must therefore be short of full width. If so, insert one end to the full depth of one slot, then slide the tread back into the other slot until each end rests halfway into each slot. Insert shims of equal thickness into the gaps where the tread ends meet the slot bottoms and secure the tread in the same manner as it was originally secured. To add support for the new tread, use cleats beneath the tread in the corners formed by the tread and the stringer. To both the stringers and the tread nail or screw one-by-ones that are as long as the depth of the tread.

If the original tread was not joined to the stringers in slots, but rested either on top of steplike cutouts in the stringers, or on cleats fixed to the insides of the stringers, your job is even easier. After removing the damaged tread, measure the width between the stringers exactly, get a replacement matching the original, and install it with the same type of fasteners that held the original.

To replace a tread in more complicated stairways, you may have to remove risers, moldings, or balusters, but the order of operations is the same. No one set of instructions can apply to all types. You must judge how difficult the job is; it may be necessary to consign it to a professional.

Repairing a Baluster

Balusters are the uprights that connect a handrail to the treads on the side of stairway not built against a wall. The designs of individual balusters are as numerous as stairways themselves, and if a replacement is needed, you may have to have the design custom-matched. Only the simplest balusters are kept in stock at most lumberyards. To achieve good quality results requires at the very least woodworking competence and care.

If a baluster has suffered a clean break, it may be possible, with the careful use of clamps, to glue it together again. A skilled woodworker or furniture repairer may even be able to join the two pieces with dowels and patches. In any event, it will help enormously to understand how the baluster is fixed in place. There are four basic methods (see Figure 21).

Simple Nailing   The crudest method, which is not uncommon, is for the baluster to be nailed at each end to the tread and the handrail. Sometimes the nailing is done discreetly from underneath, and sometimes more blatantly. In either case, however, if the baluster is in two parts, you should be able simply to pull each part off the nails, then remove the nails with pliers.

Held in Grooves   In a more sophisticated system, commonly used for balusters with square ends, the ends are fixed into a groove underneath the handrail and in the top of the tread, or in a *shoe,* a box made out of molding fixed to the top of the tread. This groove usually runs the entire length of the handrail. The spaces between one baluster and the next are filled with *fillets* (short pieces of wood) to keep the baluster aligned. Before removing the baluster, which is usually simply nailed, it may therefore be necessary to remove one or more fillets. Since the usual method is to split them out with a chisel (removing any nails as you go), the fillets will probably also have to be replaced.

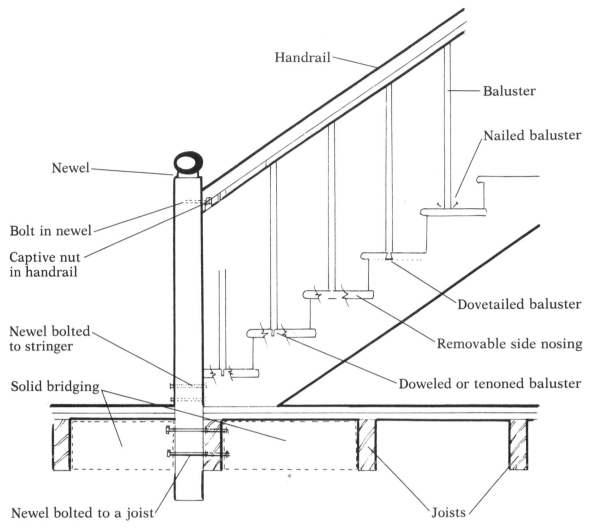

Figure 21 Ways to fix a balustrade

Mortise-and-tenon   Better built stairs have the balusters made with round *tenons* (dowel-like protrusions) at each end, which fit into corresponding round mortises (holes) in the tread and handrail. If the baluster is in two parts, each may be twisted out of each mortise with as much force as necessary, but replacement of a single piece is trickier.

First, try inserting the top end of the replacement in the underside of the handrail and lifting upward enough to slide the bottom end into its mortise. Be careful: You don't want to compromise the strength of the other balusters! If this doesn't work, shorten the top tenon a little and bevel the sides so it doesn't bind against the mortise; if you can now slip it into place, there should still be enough tenon at each end to keep the baluster fixed. As a last resort, you can saw off the bottom tenon, plug the bottom mortise (perhaps with the cut-off tenon), and fix the bottom end by careful toenailing. Again, drill pilot holes.

Dovetailed Joint   Top quality traditional stairs have balusters mortised at the top (as just described) and *dovetailed* (a wedge-shaped joint) at the bottom. The clue that the balusters are dovetailed is finding a separate piece of *nosing* (a rounded section) fixed to the side of the tread so that it covers the joint between tread and baluster.

If you carefully pry off the nosing—first removing any molding strip under the nosing— you will see the wedge-shaped end of the baluster (the dovetail *pin*) sitting in its matching cavity in the tread (the *tail*).

Hammer or chisel the pin out carefully. The replacement must be made either with a matching pin or with a round tenon that you can fit in the tail and nail into place. Predrill the nail hole to avoid the possibility of splitting the tenon. Then replace the nosing and any molding you may have removed.

Newels and Handrails

As mentioned earlier, the newel post is often an important link between the staircase and the rest of the house. Stringers may be firmly joined to the newel, which in turn is likely to be firmly secured to the framework of the house itself by being bolted to the joists and girders below the floor (see Figure 21). It is also the termination for handrails that are not otherwise fixed to a wall; thus it plays an important part in keeping the whole balustrade fixed and rigid. A good quality repair requires not only skill, but also sound engineering to ensure that the newel is secure and stairway safe.

Strengthening Newels   If a *base newel* (one at the bottom of a flight of stairs) is wobbly, see if it does go through the floor before trying to strengthen it at the floor line. If it does, make sure that the bottom end is as secure as possible, perhaps by adding solid bridging (see Tightening Springy Floors, page 17) between the joist to which the newel may be connected and its neighbor; then bolt the newel's base to the new bridging.

Now you can reinforce the newel above the floor line, if more support is necessary. Take care to keep the newel perfectly vertical. Try to place new bolts or lag bolts so that they connect the newel with the strongest parts of the staircase—the carriage is best. Drill holes and use lag bolts rather than screws if you're sure of connecting with substantial members such as heavy stringers. Plug the holes when you're done.

In good construction, the intermediate newels—at landings, turns, or at the top of balustrades—are also attached to substantial framing members of the house, such as second-floor joists. Try to add support to these members, if possible, rather than simply banging more nails through the base of the newel into floorboards. If the original construction was not done as well as it should have been, it is often impossible to secure a weakened newel safely. The only solution is to have a professional rebuild it properly.

Repairing Handrails   As well as making sure that the newel is firmly anchored to the house, you should also see that the handrail is firmly anchored to the newel. One traditional way of doing that is mortising the end of the handrail into a square section of the newel. Over the years and with use, the tenon at the end of the handrail may shrink and loosen. If it is not too loose, it can be secured by toenailing up through the handrail into the newel. Otherwise, a new doweled tenon may have to be fitted.

Another securing method involves the use of a captive nut in the handrail, a nut that engages a bolt emerging from the newel (Figure 21). If you notice a plug next to the newel on the underside of the handrail, chances are that it conceals such a captive nut. If you can remove the plug, it may be possible to tighten the nut on the bolt by placing an old screwdriver against the side of the nut and banging it around with a hammer. (Don't ruin the blade of a good screwdriver doing this.)

Many variations are possible. Some handrails do not finish at the side of a newel, but run over the top of it. It is then a simple matter to strengthen the connection with a screw down from the top, plugging the entry hole. Other handrails are simply butted up against the side of a newel and secured with screws or nails. Sometimes these are just insufficient in number or size; sometimes they just work loose. Replacement with new screws may help. A better repair is to insert lag bolts into the end of the handrail through the front of the newel. Again, plug the entry holes.

Handrails that are attached to a wall are usually held with screws or bolts into studs or framing members inside the wall. If these work loose, and larger ones cannot be used successfully, you can often reposition the brackets holding the handrail so that they can be attached to another stud. Remember that studs are commonly set 16 inches apart on center.

PART 2
WALLS

8

HOW WALLS
ARE CONSTRUCTED

This chapter will give you an understanding of the basic structure of walls commonly built for houses and other residences, so that the information about wall coverings and finishes, given in succeeding chapters in this section, can be put to better use.

Architects and builders categorize walls as either *load-bearing* or *partition* (non-load bearing) walls. For our purposes, we will divide them into *masonry* walls or *wood-framed* walls—either of which can be used for load-bearing exterior or interior walls, or non-load-bearing interior partitions.

MASONRY WALLS

Masonry walls can be made of stone, brick, concrete block, or poured concrete, or sometimes combinations of these.

Solid stone walls are relatively unusual, except in areas of the country settled long ago by European immigrants who brought a stone-building tradition with them. Many stone walls are actually framed walls covered with a face veneer of stone; for the purposes of subsequent discussion concerning any interior finish, those are treated strictly as frame walls.

Unless left bare, the interiors of solid stone walls are either plastered (and then painted or papered), paneled, or finished with a combination of paneling and plastering (see Figure 22).

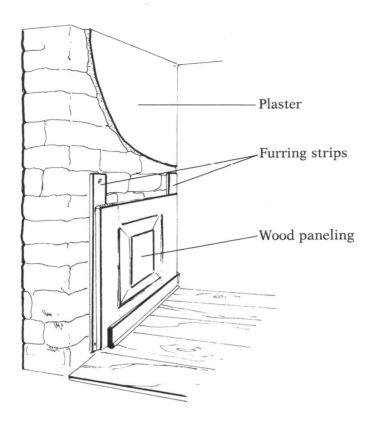

Figure 22   Solid stone wall covering

Brick walls similarly may be only a face veneer of brick over framed walls, but if solid they are invariably *rendered,* coated with a rough coat of cement or plaster, and smooth plastered on the inside. This surface is then either painted or papered.

The *concrete-block walls* most often found in residences are the basement walls, which frequently remain unfinished. Although it is possible to paint directly onto concrete block walls, most interior finishes, such as gypsum board or various forms of wood paneling, are attached on a system of wood strips (called *furring strips*) first nailed to the wall. However, it is not uncommon to find concrete walls rendered like brick walls. All of this is also true for walls made of *poured concrete.*

FRAMED WALLS

The contemporary framed wall is constructed on universally accepted principles and with standardized components. Once you understand these, you'll be able to tell with great accuracy what is inside a wall without having to remove its surface covering. Standardization was not always the case, however, and if you're dealing with an older house, it may have been built somewhat differently.

Evolution of the Two-by-four

The two-by-four is the basic unit of construction that has helped make modern framing standard. The uprights that form most of the framework of modern houses are now all made from pieces of wood that uniformly measure 1½ by 3½ inches. Despite these actual measurements, they are still known as two-by-fours, since that was the size into which such members were originally sawn.

Wood is usually sawn into various sizes as soon as the tree is felled, while it is still "green" with all its moisture. As it seasons, wood dries out and actually shrinks. Different species, seasoning under different conditions, shrink to different sizes. Before standards regarding moisture content and acceptable finished size were set, it was not uncommon to find so-called two-by-fours that measured anything but 2 by 4 inches.

Older houses not only may have been built with a variety of differently sized two-by-fours, they also may have been built without any two-by-fours at all. Prior to the introduction of powered sawmills, which made it economically feasible to produce many small timbers, houses were framed using much larger members—in a manner similar to nineteenth-century barn construction. Nevertheless, the different parts of today's standardized frame house are all descended from the basic structural members of those earlier buildings. Although the sizes of the individual parts may vary, you will still be able to understand almost any older building's construction if you are familiar with the contemporary model.

Stud Construction

Upright two-by-fours, known as *studs*, of which there are several varieties (see Figure 23), are normally placed 16 inches apart. They rest on a continuous horizontal two-by-four known as a *sole*, or *toe, plate* to distinguish it from the horizontal member that rests on top of the studs—usually made up of two two-by-fours—known as the *top plate*. The studs are placed 16 inches apart partly to make it easy to attach the sheets of wallboard, gypsum board, plywood, and other coverings that are sized in multiples of 16 inches—4-by-8-foot sheets, for example. Just as important, the 16-inch spacing has been shown to provide adequate strength under most conditions.

At corners and around windows, doors, and other openings, the upright two-by-fours are doubled or even tripled to provide additional strength and additional nailing surface. Above openings the top plate is augmented by heavier pieces known as headers, varying in size according to the distance they span and the load they have to carry.

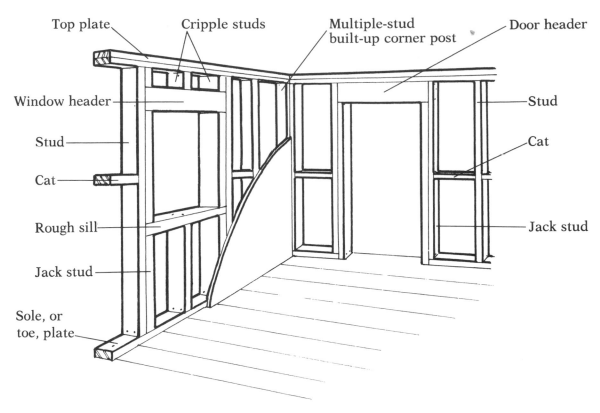

Figure 23   Standard wall framing

Strength of the Walls

The walls must be strong enough not only to bear weight but also to resist any sideways forces. This is generally achieved by *sheathing* the outside with sheet material, such as plywood, that will resist any *racking*, stresses that force the frame out of plumb. In buildings constructed without the use of plywood, sideways rigidity is achieved by the use of *diagonal braces* in the corners. One other element commonly found in framed walls are *cats*, short horizontal pieces fixed between adjacent studs. These provide extra stability and often extra nailing surface for both interior coverings and exterior sheathing.

Exterior walls are invariably load-bearing, meaning they directly support the building's weight. Some interior walls are also load-bearing walls, since they play a part in supporting the building's weight, but many are simply partition walls, whose sole purpose is to divide one room from another. When you work on a wall, unless you are going to remove, replace, or seriously modify the structure—by cutting door or window openings in it, for example— it doesn't matter much whether or not the wall is load-bearing. How the wall is prepared and finished is explained in detail in each chapter dealing with the different forms of wall covering.

INSIDE THE WALL

Regardless of their finish treatment, most walls have an inside life in common. This may include electrical wiring and various forms of plumbing. The existence of electrical wall outlets is obvious evidence that there are wires in the wall, but their absence is no guarantee that there are not. For safety's sake, always assume that there may be wires in the wall until you know absolutely that there are not. Wherever wires exist close to the surface of studs, electrical building codes require that they be protected by metal plates, but you may not find out whether this requirement has been observed until too late: Always exercise extreme caution when nailing or screwing into walls, especially in the vicinity of outlets.

Insulation is often another common ingredient inside walls, especially exterior walls, in hot climates (where air conditioning is required) as well as cold. In the normal course of events, whether or not insulation exists inside a wall doesn't have much effect on its interior finish.

9

PLASTER WALLS

ASSESSING THE CONDITION

By plaster walls, we mean that the interior wall is solid plaster, as distinct from the various forms of gypsum board, which can also be covered with plaster. We deal with gypsum board in the next chapter.

Assessing the condition of a plaster wall entails an understanding of how the plaster was applied, and over what forms of backing, since this may or not make repair worthwhile.

How Plaster Walls Have Changed

In older buildings (built before World War II) you are likely to find that the plaster has been applied over narrow strips of wood attached horizontally to the wall framing. This system is known as *lath-and-plaster,* and was for many years the standard method of plastering (see Figure 24). In very old buildings, the lath may not have been made of individual strips but cut from wider sections of wood, partly separated at alternate ends, and then stretched open to form *accordion lathing* instead.

In wood-lath-and-plaster work, each plastered area was enclosed by a border of wood *grounds,* which provided an edge for the plaster work. At the same time, it served as a nailing surface for any trim that would subsequently be applied over the edge of the plaster, such as door or window trim.

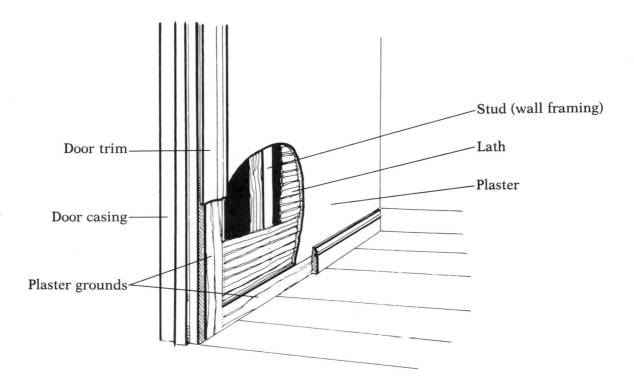

Figure 24   Wood lath-and-plaster

Around the turn of the century, wood lathing was gradually replaced by *expanded metal lath,* which became the predominant method (see Figure 25). Sheets of thin metal were stamped out in woven-type patterns and used just as the wood lath was; they were nailed to the wall framing, forming a base for the applied plaster.

A third method, also common today, is to apply plaster over perforated sheets of gypsum board, known as *rock lath* (see Figure 26), but except in special circumstances this method is inferior to metal lath.

How Damage Occurs

Plaster, when properly applied in a soundly constructed house that is kept in good condition, will last a very long time, especially if the plaster itself is also periodically maintained by painting or wallpapering. But despite its inherent strength, plaster can easily be impaired as a result of damage to the rest of the house. Excessive moisture, caused by leaks or general dampness in the house, can seriously harm plaster. Settling or movement of the house, caused by damage to the foundation or to other parts of the structure, can have a direct

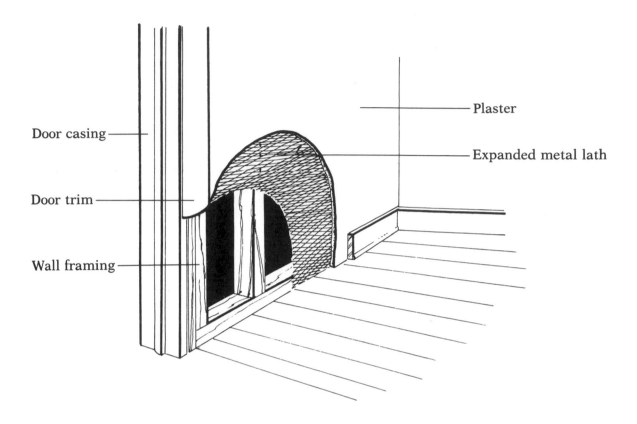

Figure 25   Expanded metal lath

and deleterious effect on plastered surfaces. Insect damage or rot in a house's framework can also cause problems.

Plastering is a specialized trade; it takes a lot of time and trouble to plaster or replaster even part of a house. What's more, it requires the skill of a professional, which is expensive. When considering repair or replacement of a plaster wall, therefore, it is important to understand what has caused the crumbling, cracking, bulging, or collapse of the plaster.

If the cause is structural movement, try to find out if it has stopped and whether the house is now stable. If the plaster appears to have suffered water damage, make sure that the cause has been eliminated before adding good work to bad. If you suspect decay because of rot or insects, remedy the situation before turning to the plaster. In most of these cases the plaster will have to be removed to cure the underlying problems. It is simply a waste of time to plaster over problems that must eventually be addressed.

The Gypsum-Board Alternative

Unless you are preserving the plaster for architectural authenticity or for the unique textured surface sometimes used with solid plaster, gypsum board can be a very suitable alternative.

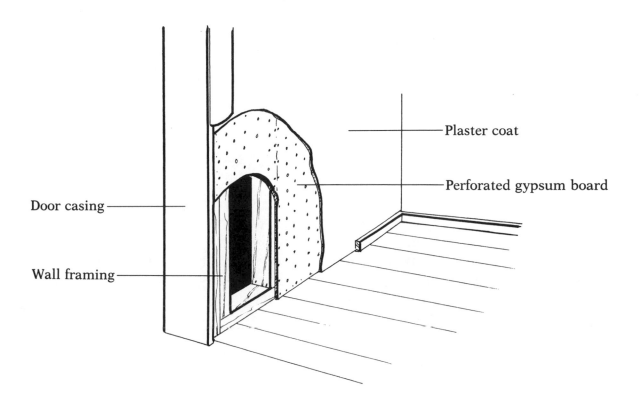

Door casing

Wall framing

Plaster coat

Perforated gypsum board

Figure 26   Rock lath—perforated gypsum board

It is not only very nearly the equal of solid plaster in many respects, but it is also decidedly easier and cheaper to apply.

If you are wallpapering the surface, no one will know whether it is solid plaster or gypsum board underneath. If you're redoing a wall surface to be painted, gypsum board is far easier to apply. While plaster is said by some to impart a more unified feeling of strength to a house, gypsum board actually has more resistance to racking forces, which tend to twist or shear a building's structure. (It is therefore considered a better interior wall treatment than plaster for houses in earthquake-prone areas.)

Plaster Moldings

Occasionally, older houses have ornate plaster moldings. Whether ceiling roses, wall friezes, or any other form of ornamental plaster casting, repair is best left to a professional, especially if they have any architectural value. It is even best to consult a professional when attempting to assess the condition of such moldings.

REPAIRS

The repairs that can be handled by the average do-it-yourselfer are limited to relatively small cracks and holes, and small areas of water damage. Anything more serious requires a professional. Furthermore, if the plaster is over wood lath and the plaster damage, no matter how slight, discloses serious damage to the lath, that should also be referred to a professional.

Repairing Cracks

Cracks in plaster that originate from the corners of doors or windows indicate that movement in the house's structure has taken place. Over many years a certain amount of settling happens in most houses and is not a cause for alarm, but if cracks continue to grow, you should consult a qualified builder or structural engineer to determine the cause and possible remedy. If the situation is stable, these cracks, and hairline cracks that may occur in other areas, can be repaired by the average do-it-yourselfer with no special tools beyond a putty knife and a broad-bladed spackling knife.

Start by removing any loose plaster. New plaster will not hold old pieces in place; it will simply be lost with the old. Hairline or other small cracks should be slightly enlarged and undercut (see Figure 16, page 69)—a soda can opener is ideal for this—to provide a base for the new material to hold on to. All loose dust and particles should be vacuumed out or blown away (keep your eyes closed if you blow the crack clean). Then dampen the crack and the area immediately surrounding it to help the new material adhere to the old. Use a small clean paintbrush dipped in clean water or a plant mister.

For small cracks the choice of filler material is either vinyl spackling compound or gypsum board joint cement—either will work fine for a small repair. Use a thin-bladed putty knife to apply the material, pressing it into the crack so that it fills it entirely. Don't put so much pressure on the knife that you crack or damage the plaster further. A small amount of the repair material should also be *feathered out*—that is, drawn out and blended into the old plaster with the putty knife.

Allow the repair to dry for a day (or longer if it is very humid); if the new material has shrunk and sunk, dampen it slightly and apply another coat. Let the second coat dry slightly, then smooth it with a wet, broad-bladed spackling knife. When it finally dries level with the surrounding plaster, sand it smooth with fine sandpaper wrapped around a small sanding block.

Patching Larger Holes

Damp or water damage can cause plaster to swell and form blisters. These are inherently unsound and should be repaired; sooner or later they will crack and the plaster will fall away. If the bellying out is not too extensive, it is best to break the bubble with light taps from a hammer and repair it as you would any other small hole: First make sure that the area to be repaired is completely free of any loose plaster, including even small pieces and plaster dust. Next, using the same kind of tool as for enlarging small cracks, undercut the edge of the remaining plaster. It should be perfectly sound and firmly attached to the wall or to whatever backing was used.

The best tool for this whole job is a broad-bladed spackling knife, somewhat narrower than the hole you are patching. Use mix-it-yourself patching plaster (available at hardware stores) to fill the hole, working from all the edges into the center. The new material is thus forced into the undercut you've made in the old plaster and is held more securely. When the whole area is roughly filled, it should be left about ¼ inch below the level of the surrounding plaster. Before it begins to dry, lightly score the surface with a pattern of crisscross lines. That makes a "tooth" to which the second coat can adhere.

In about thirty minutes, when the patch is firm but not yet completely dry, apply a second coat. It is important to mix a fresh batch of patching plaster for this second coat, since leftovers from the first batch will be too dry by now. Also, wet the area again slightly before beginning, including the area immediately around the patch. This time it will be best if you can use a broader-bladed knife wide enough so that both ends of the blade rest on the old plaster on either side of the patch at all times. This will ensure that the patch is repaired to the right level. Draw the knife from one side to the other, feathering the new material very slightly out over the old.

When this second coat has dried completely (within twenty-four hours), the patched area will be slightly lower than the original plaster. This depression can be filled with vinyl spackling compound or wallboard joint cement, feathering it out to cover the edges of the patch. After two or three days the patch should be completely dry, hard, and level; you can remove any minor imperfections by sanding lightly.

10

GYPSUM-BOARD WALLS

ASSESSING THE CONDITION

Gypsum board, or "sheetrock," as it is widely known, is probably the wall covering most commonly used in contemporary residences. Also known as *wallboard, plasterboard,* and *drywall,* it is cheap and easy to install and it provides a perfect base for painting or wallpapering. (Sheetrock is actually the trade name of U.S. Gypsum's product.) It also strengthens any structure, particularly against sideways racking forces. If it's used in sufficient thickness, it becomes a good fire barrier. Furthermore, special versions are also made with water-resistant backing paper. Unlike solid plaster, which it has virtually replaced, gypsum board can be installed much sooner in new construction, since any changes in the humidity level will not affect it as much as they would curing plaster.

What It Is and How It Is Used

Gypsum board—so called because it is made from the mineral gypsum, from which traditional plaster is also made—is actually a sheet of plaster encased by paper. Unlike traditional plaster, which is applied wet and then allowed to dry, gypsum board is already formed and dry, hence its alternative name, drywall. One side of the paper sandwich is smooth, usually cream-white, ready for painting or sizing and wallpapering; the other side, usually gray, is less smooth.

The standard gypsum-board panel measures 4 by 8 feet, but longer (up to 16 feet), as well as shorter, sizes are made. The long edges of gypsum-board sheets are made with various profiles, including square-edge, tongue-and-groove, beveled-edge, and round, but the kind most commonly used has a slightly tapered edge (see Figure 27). This creates a depression at the joint between adjacent sheets, allowing you to cover the joint with special tape and joint compound yet still preserve a perfectly flat face over the whole surface.

Gypsum board comes in several standard thicknesses, from $^3/_8$ to $^5/_8$ inch, $^1/_2$ inch being the commonest. In fact, many building codes specify that board at least $^1/_2$ inch thick be used in residences. Some codes require even greater thickness (achieved by the use of multiple sheets), since this increases the fire retardancy.

In addition to the special water-resistant gypsum board mentioned earlier, other specialty types are also available. Gypsum board with an aluminum-foil backing is known as *insulating wallboard;* another variety, know as *Type X,* is made with special fire-resistant ingredients.

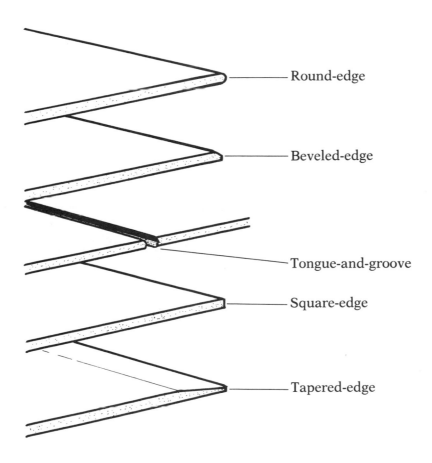

Figure 27   Long-edge treatments of gypsum board

Although most commonly applied over wood studs, gypsum wallboard may be applied over masonry walls too. Usually an intermediate system of furring strips is nailed to the wall first, to which the wallboard is then attached. Many commercial buildings use metal studs instead of wood studs, especially for partition walls. The wallboard is attached to the metal studs with self-tapping metal screws driven in with power screwguns. You are not likely to encounter metal studs unless you live in an apartment built or remodeled within the last 20 years.

Possible Damage—and Possible Fixes

Gypsum board, despite its great shear strength, is quite easily damaged. Areas not located immediately over framing members can be punctured and even have holes broken in them. Unless well protected by paint, it is also susceptible to water damage—the gypsum core is very porous and it absorbs water. Should the paper surface be torn, the gypsum, which is very crumbly and powdery, will spill out very easily.

Although dents and even quite sizeable holes can be repaired, water-damaged gypsum board is best replaced. When considering possible corrective measures for damaged or deteriorated gypsum board, the choices include not only repair or replacement, but adding another layer, since multiple layering is often an orthodox procedure at the start.

Adding an extra layer during the original installation is a different matter, however, from adding a subsequent layer later. In order to attach the new sheets securely, you must locate the framing members beneath the original material. With a wall that is properly finished and painted, you may find it extremely difficult, if not impossible, to locate the original framework. Furthermore, all trim around doors and windows, as well as baseboard molding, covers the edges of the wallboard. Even if you remove this trim, you will have a complicated job building out the wood surfaces to which the trim is attached so that these are once again flush with the surface of the new gypsum board. If the area to be covered is not too extensive or does not contain too many openings, and if you have some reliable means of locating the original studding, adding a new layer remains a viable alternative.

All things considered, gypsum wallboard is one of the easiest and most rewarding surfaces to renovate. It is surprising what a little patching and a fresh coat of paint will accomplish. If there appear to be too many patches and other repairs to be made, however, consider the possibility of total replacement. Replacing a gypsum-board wall is quite straightforward, even if it can be laborious and dusty.

REPAIRS

Aside from direct damage, such as a blow to the wall, or a swollen and damaged area resulting from a leak or serious spill (or even flooding), gypsum board can also be damaged by structural changes in a house, such as settling. This will show up in the form of popped nails or screws that should be holding the gypsum board to the studs, split joints and seams, and cracks around openings.

Once you are certain the condition that caused the damage has been corrected or has stabilized, repairs are straightforward and easy to make. They often require little more than the application of a patching material and paint. The patching material may be the joint compound that is used in the original taping of joints and corners (see page 112), or spackle, which is readily available in cans of different sizes. Ready-mix spackle is better suited for very small jobs—it tends to dry with less cracking. For hairline cracks it is easier (and cheaper) to buy joint compound, which may be obtained in 1-, 2-, or 5-gallon containers.

Redoing Popped Nails

If fasteners—nails or screws—have "popped" (poked through the surface), they may have come out as a result of shrinking or movement in the framework, or because they were poorly set in the first place. They must be set below the surface and recovered with compound (see Figure 28). Screws should be screwed further in, and nails should be driven in with a *crown-faced hammer*. (This is a regular hammer with a slightly rounded face—a flat-faced hammer might tear the paper surface.)

To prevent the nail from popping out again, you may have to drive in another nail, close to or overlapping the first one. Use only special wallboard nails or annular (ringed) nails, as they have greater resistance to being pushed or pulled out.

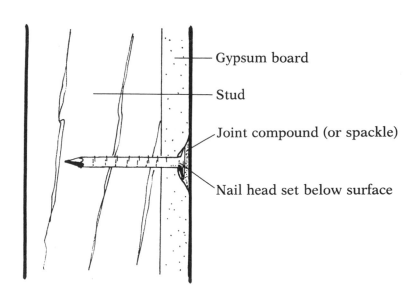

Figure 28 Properly set nail (or screw) in gypsum board

Ideally, nails or screws should not break the surface of the paper covering, since the gypsum inside is not dense enough to prevent the nail from being pulled right through it. If you are trying to nail back an area of gypsum board that has bellied away from the wall behind, don't do it one nail at a time. It is safer to push the entire area back against the framing, using a length of two-by-four, for example; then double-nail the gypsum board in place, angling a pair of nails toward each other.

The heads of all nails or screws should be set just below the surface of the surrounding board or you won't be able to cover them with joint compound and produce a perfectly flat surface. Pull a broad-bladed spackling knife across the surface to make sure that it doesn't touch the heads of any nails before starting to fill the holes.

Dealing with Small Holes and Cracks

Small perforations and ragged holes should be struck with a hammer to produce a shallow indentation that can then be filled as if it were the dimple or depression surrounding a nailhead. This is a delicate operation: Try not to hit the gypsum board so hard that you break the paper surface. You must use much less force than when you set a nail.

Filling the Holes   The procedure for filling these dimples, holes, and depressions is the same as you'd use for filling and smoothing small cracks and slightly opened seams. Using a spackling knife with a blade wider than the area to be filled, wipe some joint compound or spackle into the area. Then pull the knife across the depression, with the ends of the blade resting on the surrounding gypsum board. That will level the area. Allow the slightest amount of patching material to feather out over the edges of the repaired area.

When the patch has dried, it will have shrunk a little; repeat the process until it is perfectly flat and level with the surrounding gypsum board when it's dry. Depending on the size of the area being repaired, this may involve two, three, or even more fillings. Be patient. If you apply too much at once, it will only take longer to dry and might crack, and you'll spend longer sanding away any excess.

Sanding the Patch   Sanding is a messy and dusty process. Much of it can be avoided by careful application of the patching material in the first place. To produce an absolutely smooth finish, the material must be applied with a perfectly clean and smooth-edged spackling knife. If the knife blade is chipped or dented, or if bits of old patching material have dried into hardened nubs on the edge of the blade, you will be forever sanding the patch smooth. Keep the edge of your blade clean by wiping it on a short length of stick kept just for that purpose. Don't wipe the knife on the edge of the spackle can or compound container. The excess will harden and form lumps that may drop into and spoil the rest of the material.

Medium-size Holes

Holes too large to be filled simply with patching material require some form of backing against which to apply the wet spackle or joint compound. But when a hole is larger than 5 or 6 inches across, it's better to install a replacement piece of gypsum board (see next section).

What you need within the hole is a base against which the patching material can form and dry rather than simply being pushed unsupported into the hole. One way is to insert into the hole a piece of mesh screening larger than the hole (see Figure 29). Roll it up before you put it in and allow it to unroll when it is inside the hole; keep hold of it with a piece of string threaded through the center of it.

Before inserting the mesh into the hole, remove any loose and crumbling gypsum from the hole's edges. Moisten the edges, then cover them with patching material, pushing some of it into the hole so that it sticks to the back of the gypsum board.

After inserting the mesh, and after it has unrolled, use the string to pull it gently against the back of the gypsum board around the hole so it is at least partly seated in the patching material you just placed there. Still holding the string (or securing it to a stick that straddles the hole), apply patching material to the visible surface of the mesh, smoothing it up to the edges of the hole. When this has hardened so the mesh is firmly in place, cut off the string

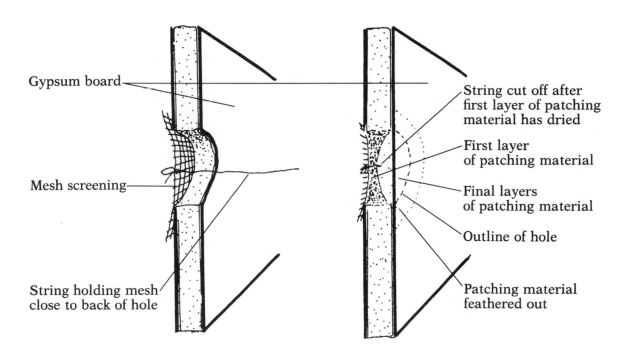

Gypsum board

Mesh screening

String holding mesh
close to back of hole

String cut off after
first layer of patching
material has dried

First layer
of patching material

Final layers
of patching material

Outline of hole

Patching material
feathered out

Figure 29   Patching a small hole with mesh

and patch the remaining hole. When this too has dried, apply more material to smooth off the patch. Use a wide-bladed knife similar to the one used for smoothing over nail dimples, described above.

Patching Large Holes

A hole or other damage greater than about 6 inches across is best repaired with a new section of gypsum board. Be sure to use matching board—the same thickness, type, and quality as the original. A rectangular replacement is the easiest shape to deal with, but since the new piece has to be well supported from behind, you can't simply use the smallest rectangle that will fit into the damaged area.

In planning the area to be replaced, you may find it necessary to cut out a much larger piece so you can align at least one of the sides of the replacement over a section of the wall framing. It has to be supported at least on two opposite sides (see Figure 30). Remember conventional wall framing generally consists of two-by-four studs no further apart than 16 inches, and that they are frequently much closer than this in corners and near window and door openings.

If, through the damaged area, you can find a stud close to one side of the hole, you should be able to extrapolate the location of its neighbor. (Use a wire to probe for a stud; if the damage is not a hole, make a hole in it—you will be replacing it anyway.) If the hole is close to the bottom of the wall, you can remove the baseboard and cut the replacement piece so its bottom edge can be nailed to the sole plate. Similarly, if you have to repair a section close to the top of the wall, you may be able to take advantage of the top plate. Finally, remember the cats (short horizontal two-by-fours between studs) are often to be found about midway up the wall. If, by locating the replacement rectangle over framing members, you are making a ridiculously large repair for a relatively small area of damage, you may have to deal with a smaller unsupported rectangle.

Removing the Damaged Section Once you decide on the area to be replaced, mark the rectangle on the wall, using a framing square and a straightedge. As noted, it must totally include the damaged area. If you've been able to position any edge over studs, plates, or cats, these edges must be centered directly over the middle of the framing members.

Use a sharp utility knife to score along the marked lines. Start lightly, dragging the knife's blade across the paper so that it cuts rather than tears. Try not to leave a ragged edge; a clean cut will be easier to cover later. Make as many cuts as necessary to penetrate the damaged gypsum board, watching out for nails when you cut along old seams. If you are cutting across an unsupported area, you can use a keyhole-style plasterboard saw, which is much faster than a knife but more likely to damage any insulation or plastic vapor barrier that might be installed behind the gypsum board. Be especially careful if you have any reason to suspect that there is wiring in the wall. If the damage extends close to any trim, remove it and include the gypsum board under the trim in the area you are removing.

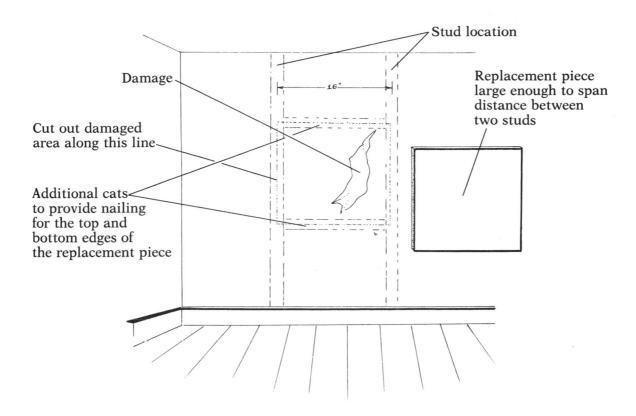

Figure 30   Repairing a large hole

Remove the damaged section and clean up the edges of the board around the rectangular hole. Remove any nails or screws that may have been exposed in a stud or other piece of framing. Check to see if any vapor barrier, including any foil-backed insulation, has been damaged; if it has, repair it by taping over the cut or torn area with waterproof tape.

Installing a Replacement Board   Before inserting the replacement piece, you have to be sure there is support for at least two opposite sides. If the patch reaches from one stud to another, this is sufficient, but if there's only one supported edge or none at all, it won't be possible to hold the replacement piece securely.

Depending on the size of the rectangle, several options are open to you. You can provide horizontal supports by securing extra pieces of two-by-four between adjacent studs, like a cat piece. You can provide vertical supports if there are nearby studs not exposed by your cutout—build out from the existing studs to the opening by nailing lengths of two-by-four to the studs (slightly longer than the height of your cutout) to create a lip of wood. A third method is to glue lengths of one-by-four behind the unsupported edges of the opening so that half their width projects; screw the old gypsum board to these pieces.

When a supporting lip has been provided on at least two opposite sides, cut the replacement piece to fit as snugly as possible without damaging the edges of the old work when it is pushed into place. A gap of ⅛ inch is acceptable. Nail or screw the edges to the support provided, taking care that the heads of all fasteners are recessed below the surface of the gypsum board. Finish the job by applying tape and compound, as described in the next section.

INSTALLING NEW GYPSUM BOARD

Installing gypsum board is hard work. Although ⅜-inch gypsum board is lighter and easier to handle than thicker varieties, local building codes may require the use of at least ½-inch board. Thicker means heavier, and also more easily damaged, since large sheets can break under their own weight if not carried or supported carefully. If at all possible, work with a helper. It is very easy to drop a sheet on a corner or dent an edge when putting it down. For a novice, a helper is essential.

Make sure that any sheets you store are stacked flat and are well supported by two-by-fours. They should be kept away from any dampness. If the supports are too far apart the sheets may sag, and if the humidity is high enough, they may "set" in this shape and be much harder to manipulate.

Planning the Layout

The essence of successful installation of drywall lies in careful planning of how the sheets will be used to cover any given surface. (*Note:* If ceilings are to be covered as well as walls, it is easier to cover the ceilings first.) Since framing is normally constructed in 16-inch increments, and since gypsum board is commonly available in 4-by-8-foot sheets, you will have the choice, in most cases, of installing the sheets on the wall either vertically or horizontally. There are advantages to both, but consider the following points before deciding.

If the ceiling is less than eight feet high, the sheets are usually installed horizontally. This results in fewer joints. However, since the sheets should be installed from the top of the wall down (to get the best joint with the ceiling), you may need a helper: it's hard to support the upper sheet and nail it at the same time. If you can't get a helper, you can use a temporary one-by-two ledger strip tacked to the studs at the right level, or a couple of supporting nails (in the studs, not in the sheet) to hold the sheet up close to the ceiling while you nail it in place.

For rooms with ceilings higher than 8 feet, installing the gypsum board vertically makes it easier to cut pieces for the remaining space above the sheet. The joints, being above eye level, are less likely to be noticed. Also, a vertical installation is much easier to do on your own.

More important than either of these two considerations, however, is the overall layout of the framing to which the gypsum board is to be attached. The position of doors and windows may dictate where it is sensible to start. That may in turn mean installing the sheets horizontally rather than vertically, regardless of the height of the room. Your aim should be to use as much of each sheet as possible, and to cover the room with the least cutting and the fewest joints.

Since it's very easy to measure incorrectly for cutouts when the electrical outlets fall in the center of sheets, give some thought—all other considerations being equal—to laying out the sheets so that the outlets fall at the edges of sheets. That also applies to cutouts for windows and doors—it's much safer to begin a cutout at the edge of a sheet than at the center.

Final Checklist

Before cutting any board, there are other items to check:

- Inspect the framing to which you will be nailing the gypsum board. Make sure that all of it is in the same plane. If one stud protrudes beyond its neighbor, it will be hard to get the gypsum board to lie flat.
- Make certain that there is a nailing surface for at least every long side, particularly in the corners and areas around openings, under windows. and at the floor and ceiling corners.
- Measure the distance that the outlet boxes protrude from the surface of the framing. Make sure they won't stick out through the finished gypsum board, or be left too far behind it.
- Inspect all wiring to see that it is routed well within the framing to be safe from the nails you will be driving through the gypsum board. If not, see that there are protective metal plates covering it.
- If trim is to be installed to cover the edge of the gypsum board where it meets doors and windows, make sure that the nailing surface for the trim will be flush with the finished surface of the gypsum board.
- Inspect the vapor barrier to make sure it is whole and undamaged. Repair it if necessary, or install one, if none exists. (Use plastic sheeting; staple it to the studs and tape the seams.)

Cutting and Nailing

When cutting sheets of gypsum board, try to save as much of the factory edge as you can. If you have to cut a piece from a sheet that has already been cut, do it, if possible, so that you are still left with one factory edge. Not only will the factory edge be straight, but if it is the long side, it will also be tapered and will help produce a better joint. Similarly, try to arrange cut edges so that they are the ones hidden under trim or behind the baseboard (see Figure 31).

Work patiently. "Measure twice, cut once" is a good carpenter's maxim to bear in mind. Make all your measurements and cuts on the face side. Use whatever aids you can: extra long straightedges, chalk lines, and very large *T squares* (designed specifically for measuring and guiding straight cuts and exact right angles in gypsum board). All these are available at hardware stores or home improvement centers.

Cutting the Board   When the board is marked, use a straightedge (even the factory edge of another piece of gypsum board will serve) to guide your utility knife as it scores the board on the face side. Take care to cut the paper; do not tear it. Change blades often.

For cuts made the full width or length of the board, fold the sheet back on itself sharply,

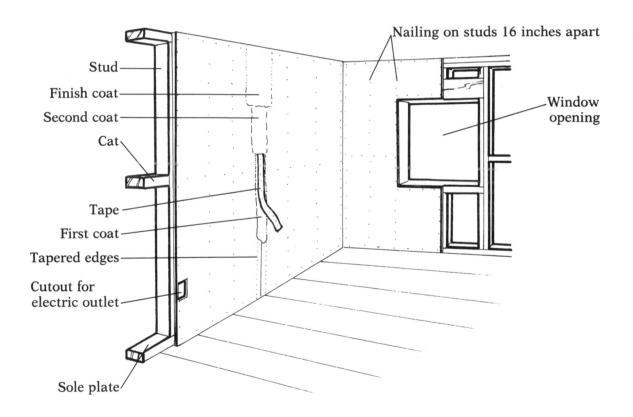

Figure 31   Nailing and taping gypsum board

away from the cut you have made. Sometimes it's easier to have the sheet standing on edge when you fold it; at other times it might be easier to work on the sheet flat on the floor. If it is on the floor, slide a two-by-four under the sheet to one side of the cut, and press down on the overhanging part. However you make the fold, the gypsum inside the paper should crack along the scored line.

The next step, with the board folded along the cut, is to score another line on the inside of the fold, thus cutting through the paper at the back. That should free the cut part so that it drops away; if it doesn't, it can easily be broken off by folding it sharply back in the other direction.

For cuts that do not extend the full width or length, you must cut completely through the board with the knife. Be patient; this can take several cuts. It may well be quicker to use a saw, but this produces a lot of gypsum dust and a very ragged edge to the paper. However, a small keyhole saw may be used to good advantage when cutting out holes for electrical outlets and switch boxes.

If you have to make curved cuts, you can score the curved line, score crosshatches in the waste area, and then break away the waste in segments. Alternately, use a saw and put up with the dust.

Nailing Up the Sheets   A helper is invaluable when nailing the sheets to the wall since even ⅜-inch gypsum board can be difficult to hold up and nail at the same time. If you have to install a vertical sheet off the floor on your own, place a small piece of wood on the floor just in front of the center of the sheet, and use another short length as a lever. Using the first piece as a fulcrum, step on the second piece to raise the board exactly where you want it. This method won't work for horizontal sheets, since the gypsum board is likely to break or be crushed at the levered point.

Use cement-coated, annular, or special plasterboard nails long enough to penetrate the studs about 1 inch and space them about every 10 inches. Use a crown-faced hammer and sink the nails enough to produce "dimples" without actually breaking the paper. What is important is that the heads of the nails all be below the surface of the gypsum board.

Taping and Cementing   When you've nailed all the gypsum board in place, and set all the nails below the surface, you're ready to finish the job. This involves filling all the nail holes, the seams, and the corners with joint compound, and then sanding them smooth. When you're done, and the wall is painted or papered, it should be impossible to see any nails or seams. This process is often called *taping,* since special perforated tape is used over the joints, along with the compound, to produce a strong and invisible seam—without it the compound would develop cracks (see Figure 31). Professionals who do this every day can obviously do a perfect job. For nonprofessionals, taping takes time and can be frustrating— do the inconspicuous joints in a room first to get practice. If you take your time, however, you too can produce very satisfactory results. You'll need an assortment of taping knives with progressively larger blades, as described below.

Filling Nail Holes   Start by filling all the nail holes, taking care to pull the blade of the spackling knife firmly across the filled hole so that none of the compound in the hole is higher than the surrounding gypsum board. If it is, you'll spend a lot of dusty time sanding it flat again after it has dried, before you can apply the next coat. Be sure you've filled all the nail holes. It will be hard to fill a missed one later if it's near a taped seam without disturbing the compound in the seam.

Taping the Joints   Next, using a medium-size *joint knife* (about 4 or 5 inches wide) and working one joint at a time, apply a generous amount of compound to the joint. Tear off a strip of tape equal to the length of the seam. Crease it down the center (it is usually slightly precreased), then open the strip out again and paste it down in the fresh compound so the crease lies directly over the seam. The crease in the tape helps align it over the seam.

Starting at one end of the tape, draw the joint knife over and along it so that the tape is forced down into the compound and flattened. If you press too hard, you will drag the tape along with the knife; if you press too lightly, you will not embed or flatten the tape enough. It helps to draw the knife toward one end of the tape for the first couple of inches to get that end firmly embedded and properly positioned before doing the full length of the tape in the other direction.

While it's important to get the tape and compound level, too many passes with the knife will damage the tape. You'll have to remove it completely before starting again. You'll quickly learn how much compound is required in the joint to set the tape. If you use too little, you'll constantly be removing the tape to apply more. Compound should spread out from both sides of the tape, but if you use too much, great dollops will accumulate on your knife and inevitably fall to the floor. Don't reuse compound that has landed on the floor; it will have picked up particles of dirt and grit that will make a smooth joint impossible. Clean it up as soon as possible; it hardens very quickly and becomes difficult to remove.

Tape in a logical progression. Start with the joints that will be crossed by longer joints later. For example, tape all the vertical joints before doing the joint where the ceiling meets the wall.

Taping the Corners   *Inside corners* are taped the same way as flat joints, except that having creased the tape, leave it folded to something greater than a right angle to make it easier to set into the corner. There is, among other specialty knives, a double-bladed corner knife. For amateurs, it may be easier to use a straight knife with an angled blade until you become practiced at applying just the right amount of compound. Holding the skewed blade so that the pointed corner is directed to the inside will allow you to run the knife down the corner without rubbing your knuckles into the wall.

Outside corners can also be taped, but it's generally better to cover them with *metal corner bead* before applying compound (see Figure 32). The use of corner bead provides a straighter edge and protects the gypsum board better at an outside corner, which is likely to suffer above-average wear. The beading is perforated to hold the compound. It is cut to length with tin snips and nailed into place, aligning the outside bead carefully to produce a straight corner. The compound should cover the entire bead up to its very edge.

Creating a Smooth Finish   When the first coat of compound has dried over the tape— usually a day later—apply the second coat. Go over all the nail dimples first, then work on the joints in the same order you did the first coat. The second coat is easier to apply using a tool with a much wider blade, such as a 10-inch joint knife. The wide knife allows you to spread the compound evenly over a wider area, leveling the compound between the surfaces of adjacent sheets of gypsum board. The second coat should completely cover the swath of compound you laid down during the original taping.

The joints formed by the *long edges* of individual sheets start out already low because of the tapered edges. This depression must be completely filled with compound, which should be feathered smoothly out at the sides. When you tape joints formed by the *ends* of sheets, it's a little more difficult to achieve a perfectly flat joint. You must pay even more

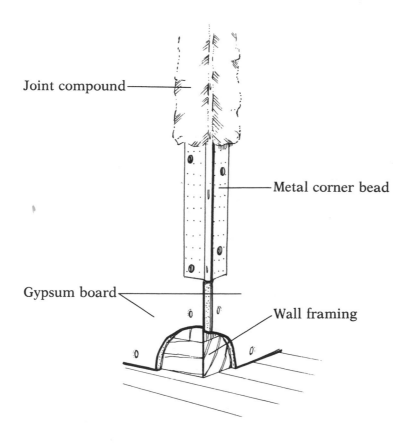

Joint compound

Metal corner bead

Gypsum board

Wall framing

Figure 32   Finishing an outside corner

attention to the feathering out of the joint compound. The first time you try this, you're likely to discover that part of the first coat was inadvertently left higher than the surrounding gypsum board. When you try to smooth the second coat, the knife scrapes against this hardened cement. The only cure is to gently sand the first coat level before reapplying the second coat.

For a completely smooth finish, it's best to apply a third coat, after the second coat has had a day to dry. Before doing so, sand any high or rough spots carefully with fine grit sandpaper. If you sand too much, though, you can easily go through the compound into the paper, producing an unfortunate hairy surface. You'll have to reapply compound to this area, allow it to dry, then very carefully sand again.

A final light sanding to remove any last nibs of compound should be all that is necessary when the last coat has dried. As an ultimate check that the wall surface is truly flat, try inspecting it with a light source held very close to it—the slightest depression will show up. You'll have one last chance to perfect it before painting or papering.

11

PAINTING WALLS—AND CEILINGS AND TRIM

Almost everyone has done some painting at one time or another. It can be one of the easiest, fastest, and most dramatic ways to change the appearance of almost any surface. Painting skills are not difficult, but they do take time to learn. Even a novice can produce very good results—with care. A purchase of good-quality brushes, rollers, and some other relatively inexpensive equipment will be money well invested toward good results.

Despite its superficial ease and simplicity, however, painting the interior walls of a house successfully calls for a certain amount of forethought and technique. Quite apart from the choice of color, which is mainly a matter of personal taste, the first stumbling block can be not knowing what kind of paint to use.

TYPES OF PAINT

Although paint has been around for centuries, with new varieties being developed constantly, there are two basic classifications into which practically all paints can be divided: *oil-based paints* and *water-based paints*. Both types are composed of a pigment—the coloring agent—mixed with a vehicle—the other ingredients that hold the pigment and keep the paint liquid until applied.

Oil-based paints are those paints whose vehicle includes natural oils, such as linseed oil, or various synthetic resins called *alkyds*. Today, because of the predominance of these new resins over the more traditional oils, this type of paint is now also commonly referred to as *alkyd paint*.

The second major type is *water-based paint*, whose vehicle is composed partly of latex—hence the common use of the term *latex paint*.

Comparing the Types

Alkyds are easier to wash than latexes and resist "blocking" better. (Blocking is the tendency of some paints to remain slightly tacky long after they are dry.) Alkyds are still the recommended paint for areas that see a lot of wear, such as interior woodwork—door trim and bookshelves, for example. On the other hand, they give off strong-smelling fumes, requiring good ventilation during application. Furthermore, they take considerably longer to dry than latex paints.

Latex paints tend to be easier to apply, easier to clean up, and quicker to dry, and so are eminently suitable for flat walls and ceilings—in fact, they are about the only flat interior paint available. Many people use latexes on trim.

Primers

Whenever you intend to paint metal or new (raw) wood, it is better to use *primer paint* first. (Some primers are occasionally called *undercoats*.) Like finish paints, they may be oil-based or water-based. The basic function of all primers is to ensure that the finish coat adheres well and covers well.

Primers are made to accommodate different finish paints on different surfaces. There are special primers for painting metal and for use on concrete, and there are primers made especially to cover stains as well. It is important to use a primer that not only does the job you want it to but also is compatible with whatever finish paint you intend to use. *Read labels carefully to make sure everything matches.*

If you are painting unpainted gypsum board or over plaster patches, use one coat of your finish paint as a primer.

The Range of Paints

The majority of oil-based and latex-based finish paints are for general interior or exterior use, and are offered in a flat finish or various degrees of glossiness. There are, however, numerous specialty paints such as epoxy paints, which are especially good on nonporous surfaces like ceramic tile, heat-resistant paints for objects like radiators and fire hoods, textured paints that hide imperfections well, and specially thick ceiling paints.

Choosing the Paint

Before you can actually paint, of course, you have to know the right kind for your particular application, how to choose between the various grades, and how much you're going to need.

Note that some paints are offered in a huge range of so-called *custom color* mixes. These are usually the same quality as factory-mixed paints, with the exact color being mixed in each can to a formula on the color card. The selected product Ratings (see the Appendix) are the best guide to determining the respective qualities of various paints. One useful tip when you're using latex paint: If it smells bad, it probably is! Some of the ingredients of latex paints can actually rot. Do not use paint that smells like curdled milk or paint that has been frozen.

How Much You'll Need

The easiest way to estimate how much you'll need is to calculate the rough surface area: Measure the perimeter of the room (or the length of the wall, if that's all you're painting) and multiply that by the height of the walls. That will give you the square feet you need to cover.

You can calculate the actual square footage more exactly by subtracting 20 square feet for every door and 15 square feet for every window, for example, but it's better to allow some extra.

Different paints cover more—or less—square footage, and different surfaces may absorb different amounts of paint. Also, some paints hide (cover a darker color) better than others. Therefore, estimating quantity is not an exact science; adjust the amount you buy to your situation. Most labels say the paint should cover 400 to 450 square feet, but on the average 650 square feet is closer to the mark. However, there is generally no such thing as one-coat coverage; most jobs take at least two coats. Of course, it almost always takes more coats to cover a darker color with a lighter one. Finally, you should always have more than you need (in case of unexpected spills or subsequent repairs), and if you go back later for more paint, you may not get exactly the same batch or the same mix with exactly the same color. Consider buying extra paint in smaller amounts so that you can return unopened cans for credit. (Custom-mixed colors generally cannot be returned.) Check to make sure the supplier will accept returns.

BASIC PAINTING

Preparing the Surface

Even the correct type of paint, no matter how carefully applied, can fail if the surface is not properly prepared. Do not skimp this essential step. Professional painters commonly estimate they will spend one-third of their time in preparing the surface. (They don't want to be called back if the paint starts wrinkling, blistering, or peeling.) A sound surface is the first requirement of a good paint job.

All holes, cracks, and damaged areas of plaster walls or walls covered with gypsum board should be repaired, as described in the previous two chapters. Any damaged woodwork should be repaired too. Paint may cover a multitude of surface blemishes, but it will not fill gaps or mend cracks. "Spot prime" the patches with one coat of the finish paint you plan to use.

Wallpapered Walls   If the wall to be painted has previously been wallpapered, see if the paper is a self-stripping type—tug on a corner and if it comes away, remove all the paper. If the paper doesn't self-strip, it may still be a good idea to remove the paper by soaking it. The paint will stay on the wall only if the paper does; if the paper subsequently starts to peel, you will have wasted paint, time, and money. (See Removing Old Wallpaper, page 130.) If you do decide to paint over the paper, glue down any loose spots and slit and glue any blisters. If any paper does come off in patches, feather those depressed areas level with patching compound. As a precaution when painting over wallpaper, prime the wall with an oil-based primer, especially if the paper is one with a very strong pattern.

Woodwork   Previously painted woodwork may present some problems. Any paint that is badly peeling needs to be scraped and sanded. Spots that are blistered or peeling can be scraped until there is no more loose paint, then leveled with patching compound. Finally, the entire surface should be lightly sanded, especially if the old paint was glossy, to help the new paint adhere properly. There are also various deglossing agents that provide a "tooth" for the new paint. All should be used strictly according to the directions.

Preparing to Paint

Finally, when all surfaces to be painted have been rendered smooth and sound, make sure that all surrounding areas are also spotlessly clean. If you've sanded, be sure to remove all traces of dust, since paint will not adhere to a dusty surface. Use a vacuum cleaner and wipe all surfaces with a slightly damp rag, but make sure that when you have finished, everything is dry as well as clean.

Before opening the paint can, take a moment to look around.

- Is everything clean? Do you have a tack cloth (see below) to clean trim and clean rags and solvent to wipe up spills?
- Have you removed as much furniture as possible?
- Is the rest covered with dust sheets or plastic and positioned so you will be able to work around it as easily as possible?
- Is the floor protected with newspaper, plastic, or drop cloths, and are nonremovable light fixtures encased in plastic bags?

- Have you taken off all outlet and switch covers—taping their holding screws to the plates?
- Have you removed or protected other trim that you don't want painted or spattered, such as doorknobs and push plates?

Last of all, inspect the surface to be painted once more for any cracks or holes you might have missed. When using any primer or paint, including latex-base types, always ventilate the room well: Establish cross ventilation with open doors and windows and a window fan set to exhaust.

Mixing New Paint

Most paint stores will have shaken or mixed your paint for you at the time of purchase, and if you open the can soon after, a light stirring is all that should be necessary. If the can has been standing for a while, however, it may need some serious stirring before it is ready to use. Use a paint stick to stir rather than a power stirrer, especially with latex paints. Too vigorous an agitation tends to produce air bubbles that get trapped in the paint. In the case of latexes, the bubbles can be transferred to the wall and blemish the job.

The best way to prepare paint and make sure that the pigment, the binder, and the thinner are all mixed thoroughly together is to *box* the paint (see Figure 33). When you open a can, if you can see separate layers of liquid—thin on top and thicker below—pour off the thinner liquid into another container (a small plastic or metal bucket, available at paint stores, is ideal). With a clean stick, stir the remaining liquid until it is of an even consistency, with no lumps. Now pour the thin liquid back into the can, stirring thoroughly the whole time. Continue to pour the paint back and forth between the two containers, stirring well each time you add the paint to the original can, until the color is uniform and displays no light streaks.

Straining Old Paint

If the can you open has already been used, a *skin* (a stiff top layer) may have formed over the remaining paint. Don't try to mix in the skin; remove it as carefully as you can and dispose of it. The remaining paint must now be strained so that you can be absolutely sure it has no lumps. Simply stretch cheesecloth (obtainable at paint or hardware stores) or an old nylon stocking over the top of another clean can or bucket, and slowly pour the paint through it (see Figure 33). Make sure that the now empty original can is free of any loose bits of old dried paint or dirt, remove the straining cloth, and pour the paint back again.

Skin is formed by the action of the air (in the empty space in a partly used can) on the remaining paint's surface. This action of the air on the surface of the paint is, after all, how paint is designed to dry. The condition in the can, however, can be aggravated by a lid that's not really tight, which allows a constant supply of air to enter the can.

The lesson should be clear—always replace the lid tightly. The main reason it's difficult

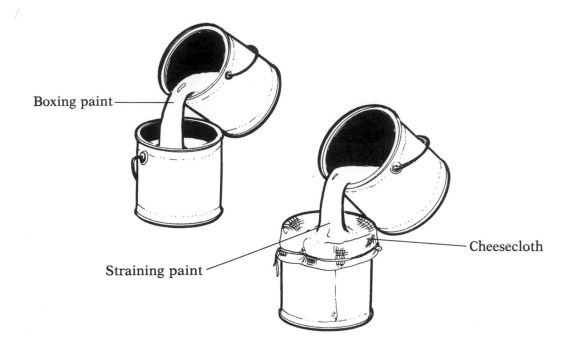

Boxing paint

Cheesecloth

Straining paint

Figure 33   Preparing paint

is that the lip around the edge of the can becomes filled with paint. Therefore, when painting, keep this lip clean—don't drag your brush across it. If it does fill up, use the corner of your brush to clean it out periodically, before the paint in it dries. Don't punch holes in the lip to drain the trapped paint back into the can (a common practice)—these holes will also allow air to enter the can when the lid is replaced.

Thinning Paint

Never thin a water-based (latex) paint. But after you've boxed, stirred, and strained an oil or alkyd paint (if necessary), it may still not brush out or roll on perfectly. The paint may seem to pull away from the wall back onto the brush or roller, the grooves left by the bristles may not disappear and flow into one another, or the paint may simply seem too thick. You may have to thin it.

First check that you have the right brush or roller for the type of paint you are using. You'll usually find recommendations on both the paint cans and on the brush and roller packaging. Then check the paint can—it should tell you whether the paint can be thinned and, if so, what to use as a thinning agent. Generally, paint thinner is used for alkyds. Follow the instructions. Add a little at a time, stirring well as you do so. Remember: You can always add more; you can't remove what you've added.

Applying the Paint

Professional painters use sprayers to cover large areas. You can rent air-atomized spray equipment, but it is probably more trouble than it's worth. Spraying may or may not be faster, and there are drawbacks.

Air-atomized spray equipment uses compressed air to force a mixed air-and-paint aerosol out a nozzle. The equipment requires a lot more attention, and can be potentially dangerous if not used with care. Also, there is invariably a certain amount of overspray—it's difficult to keep the paint only on those surfaces you want painted. Everything else must be completely masked and protected. We recommend you avoid using air-atomized spray equipment.

Moreover, *airless spray guns* are considered very hazardous. They force paint droplets— unmixed with air—out of a nozzle at extremely high pressure. If the operator should accidentally pull the sprayer's trigger while *any* part of the body is touching the opening, he or she will probably be injected with paint, causing tissue damage, quite possibly infection, and perhaps the need for amputation.

Techniques with a Brush   Though the easiest way to paint large wall areas is with a roller, you'll still have to use a brush for corners, edges, and other places that a roller can't reach (see Figure 34). In fact, the usual procedure when painting a room is to start by *cutting in*

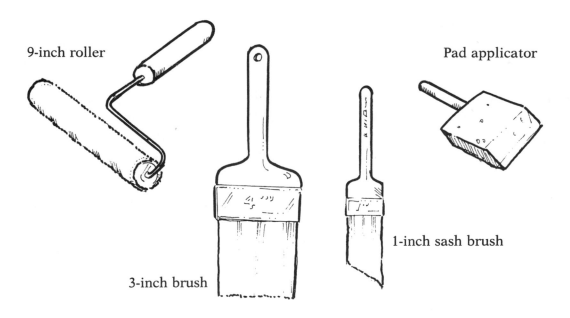

9-inch roller

Pad applicator

3-inch brush

1-inch sash brush

Figure 34   Paint applicators

with a brush. That means painting the corners and edges first with a brush, then filling in the remaining area with a roller. Although a wider brush will lay on more paint per stroke than a narrower brush, the size you use should be determined in part by how skillful you are. Most do-it-yourselfers should use no more than a 3-inch brush to cut in.

As a rule, work from the top of the room to the bottom. Here's the order for painting a room:

1. Cut in with a brush all around the ceiling, then paint it with a roller.
2. Cut in with a brush around the top of the walls, all room corners, and around doors, windows, and other openings.
3. Paint the walls with a roller.
4. Paint the trim with a brush.

When painting trim, work again from top to bottom: picture molding (if any), door trim, and windows, and then wainscoting (if any), baseboard, and finally doors themselves last. When painting windows and door openings, work from the inside surfaces out to the trim.

Here's the basic technique for loading a brush with paint:

1. Dip the brush in the paint no more than halfway up the bristles.
2. Gently tap the brush against the side of the can, first on one side of the bristles, then on the other (do not wipe the brush over the lip).

There's no point in filling the brush with paint if you wipe most of it off again; your aim is to fill the brush as full as possible but still be able to remove it from the can without dripping paint. Tapping the brush against the rim will accomplish this.

When painting windows, doors, and trim, wipe all surfaces with a *tack cloth* just before painting them—this removes all traces of dust. Tack cloths are generally cheesecloth impregnated with substances to make them sticky; hardware and paint supply stores sell them. Using a tack cloth makes painting easier and improves the quality of the job.

For most paints it's best to use light, long strokes; the paint will not even out well if you brush it around excessively. Start with an upward stroke if possible—that produces fewer drips. As mentioned, work from the top to the bottom—that should take care of any drips that may have occurred. If an area is painted unevenly or in separate sections, it may show all kinds of overlapping lines when dry. To avoid that, paint into the wet edge of the previously painted section; paint continuously, finishing each new section by feathering into the edge of the previous section while it is still wet. Feather into the previous section by smoothly and gently lifting the tips of the brush as you draw paint into it.

Tips for Painting Neatly   For painting neatly in difficult places, there are specially shaped brushes (Figure 34). Relatively narrow brushes with an angled edge, known as *sash brushes*, are good for window sashes (hence their name), but they are also very useful for painting corners. The angled edge enables you to point the sharp end with precise control, and avoids getting paint on the adjacent areas. There are also very narrow brushes to get into hard-to-reach areas.

When trying to paint neatly to a line, place the brush on the surface to be painted a few inches away from the line. Then, bending the bristles so they point toward the line, carefully push the brush up to the line. When the tips of the bristles are on the line, pull the brush along, keeping the ends of the bristles on the line and using the whole width of the brush rather than just the corner.

Of course, you can also take the time to mask all adjacent areas with masking tape, and just paint right over the edge, removing the masking material after you have painted. Special smooth-surfaced tape such as Scotch Fine Line Tape works best. (If you do this, remove the masking *before* the paint dries.) Masking is a good technique for painting around window glass, but if you can exercise a little more care, it is much quicker to paint neatly using a *painting guide*. This can be any flat and thin but stiff material, with a good straight edge. It is held close to the work, protecting adjacent areas from the brush as you work (see Figure 35). You must keep the guide clean so that any paint that builds up on it does not eventually leak around the edge into the area you're protecting.

Techniques with a Roller   Painting with a roller is much faster than brushing, but rollers still should be used with care to produce an evenly painted surface without spattering the surroundings. Use the recommended roller (or roller cover) for the type of paint you are

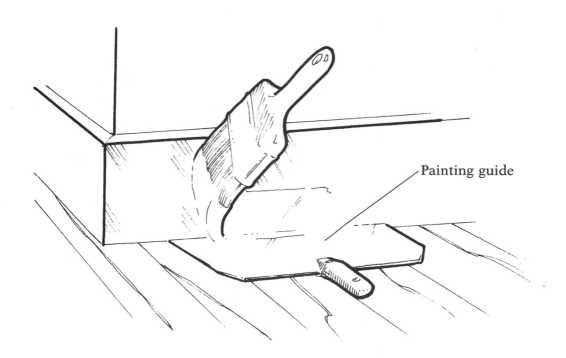

Painting guide

Figure 35   Using a painting guide

123

applying. There are different size rollers available, but the narrowest ones are not worth the trouble—it would be easier and better to use a brush since, despite its speed, a roller rarely lays paint down as smoothly as a brush.

Fill the bottom of the roller tray with paint and load the roller: Work the paint thoroughly into the roller's nap by dipping it lightly into the paint and rolling it back and forth on the gridded flat section of the tray. If the roller doesn't drip when you lift it out of the tray, it is ready to use. Don't immerse the roller completely in the paint when you are loading it— you want paint only in the nap, not inside the roller or dripping down the handle.

Just as with a brush, work from the top of the wall *down*. To minimize spattering, start each stroke of the roller gently and in an upward direction. Paint up-and-down zigzags first, then side-to-side zigzags, covering the first pattern.

As you go, increase the pressure on the roller as it empties in order to maintain a constant flow of paint. Paint sections no larger than can be completely covered with one roller load, and always work from the new section *back* into the previously painted section, lifting the roller while it is still moving as you cross the edge of the previously painted area. This feathering technique will help avoid obvious lines at the junctions of separately painted sections.

Cleaning Up

Although it may seem distracting to lay down your brush or roller, cleaning up drops and spatters as you go will save you time in the long run. Wet paint is much easier to remove than dried paint. Furthermore, it will also save you from compounding any mess by stepping in wet paint. Equip yourself with plenty of clean rags and the appropriate solvent (or water, if you are using latex) before starting to paint.

While waiting for the first coat to dry, be sure that all cans are tightly closed, and protect your brushes and rollers from becoming encrusted with dried paint. You don't have to clean them completely at this point. You can temporarily store the brushes and rollers by wrapping them in aluminum foil or keeping them immersed in water (if you have been using latex) or in paint thinner (if you have been using alkyd). Don't stand a brush on its bristles—you can suspend it in the liquid by hanging it from its handle. Drill a hole in the base of the handle just above the bristles, insert a nail or thin rod through it, and rest the ends of the rod or nail on the edges of the container (see Figure 36). There are also commercial gadgets you can buy to do the same job.

After the Job   When the painting is completely over, close tightly any cans containing leftover paint. Make sure the lip is clean, then lay a rag over the can to prevent splashes and with a hammer tap around the entire perimeter of the lid. Label each can clearly with the color, the area or object for which it was used, and when it was used. If the can is more than half empty, you will minimize the formation of any skin on the surface of the remaining paint if you transfer the paint to a smaller container with less remaining air space. Or turn the tightly sealed paint can upside down. Then store upright. The paint itself will have sealed the lid. All paints should be stored at room temperature.

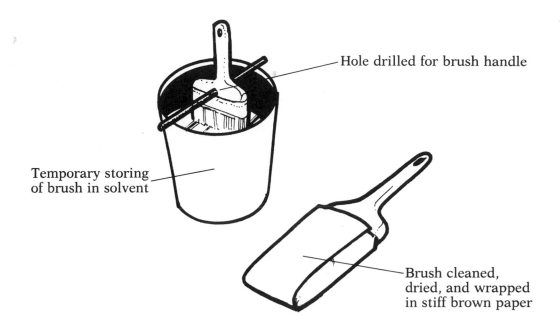

Hole drilled for brush handle

Temporary storing of brush in solvent

Brush cleaned, dried, and wrapped in stiff brown paper

Figure 36 Brush care

Cleaning paint tools is a nuisance. While it is fairly easy to clean tools used with latex, which is water soluble, do so quickly, before the latex dries. Paint thinners to clean up alkyd can be quite toxic to use and hazardous to dispose of. When using solvents, cross ventilate the room. On the other hand, it rarely pays to use poor quality, throwaway brushes; it will probably always be worthwhile to carefully clean and store a good brush.

To clean a brush thoroughly, brush out as much remaining paint as possible on newspapers, then allow the brush to soak for a few minutes in the proper thinner. Work the brush back and forth in the solvent, forcing the solvent well up into the base of the brush. Change the solvent regularly, until it remains clear, then shake out the excess solvent. (If you can find a protected area, or a large enough empty box to do it in, twirling the brush handle between two hands will spin off most of the liquid.)

Wash a brush for latex paint one final time in warm water and detergent (but not brushes with natural bristles, which may swell). Dry the brush again by shaking or spinning, straighten out the bristles or work a *brush comb* through them, and wrap them in heavy paper so that the proper shape is maintained (see Figure 36). Either hang them up or lay them flat. Finally, properly dispose of any used paint thinner; don't simply pour it down the drain, throw it out with the trash, or dump it into the ground. Call your municipal department responsible for trash pickup or your county or state environmental protection agency to seek advice about how and where to dispose of solvents properly.

12

WALLPAPERING WALLS

CHOOSING AND ESTIMATING

Types of Wall Covering

Walls may be covered with an astonishing variety of materials, including paper, vinyl, metallic foil, various fabrics, cork, and a whole array of exotic materials such as strawpaper, grasscloth, suede, and even animal skins. The three types most commonly used, however, are paper, vinyl, and polyester or aluminum foil. The technique for hanging each of these varies slightly, but in general the application is similar enough that most of the information in this chapter will apply to all three.

Wallpapers   Wallpaper is made in the widest range of qualities and designs. Standard, *machine-printed paper* is the most widely used. It comes in widths ranging from 15 to 54 inches, but the most usual widths made in America are 18 and 27 inches. This paper is nearly always sold pretrimmed and prepasted. *Hand-screened paper,* which is much more expensive, is nearly always sold untrimmed (still with its border—called *selvage*—attached). It must be trimmed before being hung. Regular papers are porous and can be hung using premixed or wheat paste.

Vinyl Wall Coverings   Some papers are given a thin film of vinyl to increase their washability, but these are not as dirt resistant and stain resistant as all-vinyl coverings. Genuine *vinyl wall coverings*—sometimes called *vinyl papers*—are the most durable and easy-to-clean coverings of all. Available with cloth or paper backings, they are smoother than vinyl-coated papers and do not tear as easily. But since they are not porous, special mildew-resistant adhesives (not wheat paste) must be used, especially when they are hung—as is most common—in moisture-laden areas like kitchens and bathrooms.

Foil Papers   A third group of wall coverings is manufactured from aluminum or polyester film. These *foil papers* may be either paper-backed or vinyl-backed, and are very reflective. That makes them especially useful in rooms that can benefit from as much reflected light as possible. Foil papers, like vinyls, are not porous. They too must be applied with mildew-resistant adhesives.

Foil papers are more difficult to hang since every tiny imperfection will show; they should generally be used only over backing- or lining-papers. They should not be hung over old wallpaper, since, being impervious, they will trap moisture underneath. The moisture is likely to loosen previously applied paper. Have foil paper pretrimmed, if you can. The papers are so easily damaged that removing the selvage can easily result in permanent wrinkles. Finally, be careful of metallic foil around electrical outlets—it may be conductive.

Flocked Papers   *Flocked paper,* a type popular for hundreds of years, was originally made only on paper but is now made also on vinyl and foil. The raised designs, formerly made of powdered wool but now made also of nylon and rayon, give these papers a very luxurious look. Like any paper with an embossed (raised) design, flocked paper must be handled carefully during hanging so designs will not be damaged. They are better suited to formal living rooms and bedrooms rather than to heavily used family rooms and children's rooms.

Picking a Paper

The choice of colors and patterns of wall coverings is so wide that unless you know exactly what you want, it would be wise to visit more than one dealer when looking for ideas.

One important point to watch for, however, is the *repeat* of the pattern. Many papers are designed with a regular pattern that must be aligned with the same part of the design in adjacent sheets. Such papers are known as *repeat-pattern* or *repeating-pattern papers* (see Figure 37).

There are two basic systems of repeats. The first, known as a *straight-match repeat,* allows a straightforward side-to-side match. The pattern is so arranged that it repeats across the width of the sheet. The second, known as a *drop-match repeat,* involves a pattern that repeats at regular vertical intervals. To align adjacent sheets of such paper properly, you have to allow an interval that varies from a few inches to as much as several feet. The importance of this becomes apparent when you have to estimate how many rolls will be needed for a given area.

Bear in mind as you shop that the wider the paper, the more difficult it is for the amateur to hang it.

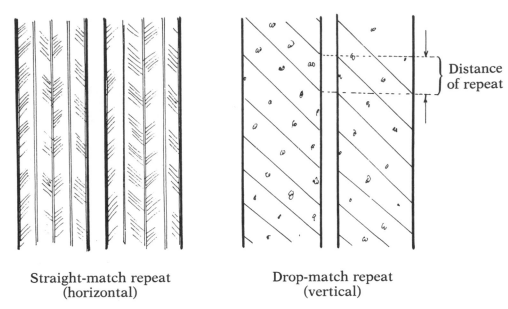

Straight-match repeat
(horizontal)

Drop-match repeat
(vertical)

Figure 37   Repeat-pattern wallpapers

Estimating

To estimate reliably how much paper you'll need for any given area, you need to know two things: first, how paper is measured, and second, the square footage of the area to be covered. The second is relatively easy; the first is not.

To begin with, all wall covering is made in *runs*. A strip of paper up to several thousand feet long, printed from one batch of dye, is one run. Since batches of dye are mixed for each run, the subsequent runs may or may not exactly match the first run. *It is important, therefore, to buy all you need at the same time.* This is the only way of guaranteeing that all the paper will be from the same run and will therefore match exactly. Estimating correctly, as you can see, becomes crucial.

At the store, wall coverings are commonly sold in *bolts,* selling units that usually contain two, but sometimes three or even four, single rolls of paper. Although wall coverings may vary greatly in width, a single so-called American roll commonly contains 36 square feet of paper. On the other hand, a single so-called European roll (generally the more expensive, designer-made, and hand-screened papers) contains approximately 28 square feet of paper.

Rough Calculations   Armed with these facts, you can make some rough estimates of how many rolls you'll need based on the square footage of the wall space to be covered. (To compute the square footage, simply multiply the combined length of the walls to be covered by the height of the ceiling.) The rough number of rolls you'll need will be the total square footage divided by 36 (if you're using American rolls) or 28 (if you're using European rolls). This will be a rough *estimate* only, since there are other factors involved that may vary from situation to situation.

Although the standard American roll contains 36 square feet, you probably won't be able to use more than 30 square feet of the roll. You'll have to trim the ends off several *strips,* the individual lengths that run from floor to ceiling. You'll lose square footage at cutouts and halves of strips to accommodate out-of-plumb corners and other inconsistencies in the room. Similarly, count on being able to use only an average of 22 square feet out of a European roll's nominal 28 square feet.

On the other hand, this rough estimate does not take into account openings where no paper is needed, such as doors, archways, and windows. That could alter your estimate substantially. What's more, your rough estimate made no allowance for the effect of repeat-patterns, which always demand more—sometimes only a little but other times a lot.

In view of all these variables, it might be best to take the measurements of the room and its openings to your dealer, and trust in his or her estimate, depending on what paper you have chosen. Also, ask if you can return an untouched bolt, perhaps for a small charge. If so, you can feel freer to order extra paper.

Estimating for Drop-Match Papers   If you elect to use an expensive drop-match repeat paper, however, buying paper on the basis of a rough estimate could turn out to cost a lot more than it should. You should estimate drop-match quantities more exactly—here's how it's done:

Measure the exact distance (in inches) between the repeated elements and divide the height of the wall (also in inches) by this distance. This will give you the number of vertical repeats that will occur in any one strip. This figure will probably include a fraction; round it up to the next highest whole number. Now multiply the distance between repeats by this whole number. The result will be the adjusted height (in inches) that you should use as the wall height when calculating the room's square footage.

Next, calculate the area of all the openings and subtract this total from the room's total square footage. Divide this figure by 30 (for American rolls) or 22 (for European rolls), and you will now have a much closer estimate of how much paper you need. To arrive at an even more accurate estimate, plan exactly where the strips will go on the walls (you'll have to do this anyway before starting to hang the paper—see page 131). You'll be able to see where more or less paper will be required.

To minimize the amount of waste that occurs at the end of each roll, order by bolts of two or three rolls rather than by single rolls, unless that forces you to order extra rolls.

PREPARATION AND APPLICATION

Preparing the Walls

Hanging paper requires careful planning and execution, but it can be undertaken by anyone willing to exercise that care. For all but the most experienced do-it-yourselfers, a helper will provide an invaluable extra pair of hands: work with another person if at all possible. Rooms with many corners, unusual openings, or several fixtures—such as bathrooms or kitchens—are more difficult to paper. You may want to practice on a back hall before undertaking those projects. You'll need to purchase or rent some special tools and equipment, as discussed below.

Any wall to be papered should be clean, dry, smooth, sealed, and secure. Basic repair of cracks and other damage should be taken care of, as described for painted walls in the previous chapter. The surface should contain no bumps, lumps, or other features that will show through the wallpaper.

Removing Old Wallpaper   If the wall is already papered, it's best to remove the old paper rather than applying new paper over it. If the old paper is firmly attached, it may be sound, but there is no guarantee it won't lift in the future. Papering over vinyl paper is especially difficult unless the surface is sanded or deglossed well (the same applies to foil and plastic-filmed papers), but vinyl papers are often easy to peel off. If you can strip off the old paper, wash the walls with warm soapy water to remove all traces of the old glue, then rinse them to remove all traces of the soap.

For paper that does not peel off easily, there are other means of attack. You may be able to obtain a chemical remover and rent a spray applicator from your wallpaper dealer. If so, follow all precautions and wear rubber gloves and safety goggles. When the chemical solution has been sprayed onto the paper, strip the paper off using a broad-bladed spackle knife, but take care not to nick or scratch the underlying surface. When the paper is off, wash down the walls to remove all traces of the chemical, then use a sealer on them.

You can't use a chemical remover on nonporous paper, nor will it work well on walls with more than one layer of paper. In such cases, the best solution is a steamer, which will soften the adhesive sufficiently to allow the paper to be stripped away. Steamers can usually be rented from wallpaper dealers and are quite safe to use. Work from the top down, a strip at a time. It may help to perforate the paper first, to allow the steam to penetrate better, especially at the seams.

When faced with paper that can't be removed, simply prime it with an oil-based primer, first making sure that all the paper is firmly attached. Any small loose spots should be glued down; if you remove any paper, that area should be leveled with patching material such as spackle or joint compound.

Sealing the Wall   Since wallpaper is often susceptible to stains and discolorations from alkalies or other corrosive elements leaching through to the surface, the wall to be papered should be sealed. An oil-based primer is a good sealer and will also improve the paper's adhesion—it reduces the wall's absorption of moisture from the paste. Walls that are already painted should be washed, and if glossy, made less so—either by using a deglossing agent or by light sanding.

To make sure the wall is well sealed, and also to facilitate future removal of the paper, it's a good idea to give the walls a coat of *sizing*. This is simply a coat of thin wallpaper paste, applied and allowed to dry completely. Special sizing paste is sold in premixed liquid or in powder form. Be sure to use a vinyl paste if there are conditions conducive to mildew.

The Final Steps   If you're putting up a special paper that is particularly sensitive to irregularities in the wall surface, cover the walls with *lining paper*. This is a plain wallpaper that may be hung as convenient; it does not have to be matched or even butted tightly strip to strip or against the trim. So long as its edges are within ⅛ of each other and any trim, that will be sufficient. If it is easier to hang it sideways, do so. (Paper hung sideways, above or below windows or other openings, is known as a header, not a "strip.")

When you are sure the wall has been properly prepared (this includes removal of all outlet and switch plates and any wall-mounted light fixtures), provide yourself with a trimming surface. You can often rent a professional table and other tools from a wallpaper store or rental center. Any table about 2 feet wide by 6 feet long will usually be sufficient, but make sure it doesn't interfere with your reaching all parts of the room. If you don't rent tools, buy yourself a kit, since the specialized cutter, roller, and brush are necessary. The kit usually has a chalk line, too. You'll need a bucket and sponge to wipe off excess, and, if using prepasted paper, a special water box to wet it.

Laying Out the Room

Even if you're using a solid color, nonpatterned wallpaper, it is still important to calculate the exact position of every strip on the wall. You will almost always want to hang all the strips perfectly vertical, not end up with various strips slanting in different directions.

With patterned paper, any mismatch between adjacent sheets can be quite obvious, but since few rooms will accommodate an exact number of full-width strips, you'll usually have at least one nonmatching seam, if not several. The problem is to decide where they should fall.

The main objective should be to prevent the mismatches from being too obvious. One way to do this is to locate full strips and perfect matches at the most obvious spots in the room—the wall opposite the doorway or the wall faced by the main seating, for example—and let the mismatches occur in more obscure parts of the room—over high doorways, in corners, at the sides of bookcases, or on the side of protrusions or in alcoves. Each room has its own specific problems and will demand its own compromises. But by planning the

131

layout of the whole room before you start, you won't be faced with a glaring (and un-avoidable) mismatch in the middle of the job.

Starting Points   If the focal point of the room—where you will want perfect matches—is an entire wall with no openings, you have several choices. You can start in either corner, and hope that the less-than-full-width strip will not be too noticeable at the other end. Or you can locate a seam at the center of the wall and work out to each end (thus equalizing the narrow widths at both ends). Or you can locate a strip so that it is perfectly centered in the wall (see Figure 38). The last two are generally preferable; choose the one that results in the odd section at either end being at least 6 inches wide. Strips narrower than this are difficult to deal with and, depending on the pattern, may look strange.

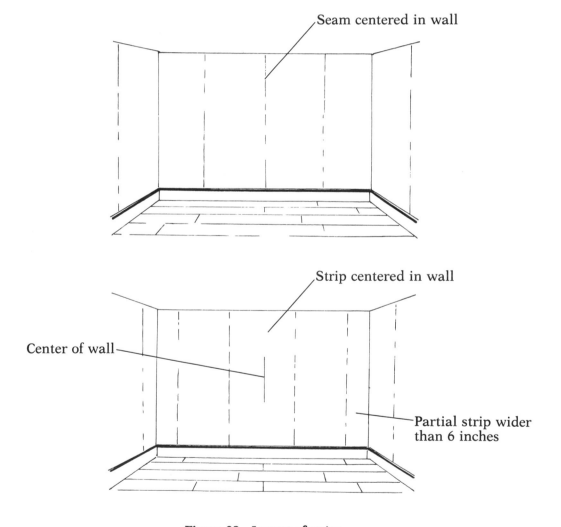

Figure 38   Layout of strips

Strips (or seams) should similarly be centered over fireplaces, and between windows, in bays, alcoves, or on chimney breasts, especially if these occupy prominent locations in the room. Remember, you want to center the paper, either by seam or strip, without leaving a strip narrower than 6 inches at either side. Occasionally, this may not be possible; if not, try to position the narrow strip in the least noticeable place.

Even after you have arranged for centered widths, there may still be a problem of matching patterned edges. Arrange such junctions so they occur in corners rather than between corners. An acceptable alternative is to locate the mismatch at the short seam above the corner of a door or a window.

Plumb Lines and Final Layout Decision   Unlike painting, wallpapering does not have to be done all at once, nor in any particular sequence. Nevertheless, for a good job, part of the layout process should include establishing the order in which the strips will be hung.

When you have completed the hanging plan for the whole room, mark where the seams will fall (short lines made with a spirit level are sufficient). When actually hanging the paper, you will work from the focal point out to the corners of that wall, then, usually, around the entire room from each corner toward the most obscure corner, where there will almost certainly be a mismatch of pattern. The short, marked lines will remind you of your layout plan.

Before you start hanging, strike a plumb line (an exact vertical that runs from ceiling to floor or top of baseboard) ¼ inch to the side of the mark in the middle of the wall where the first strip of paper will be hung. Use a *plumb bob* (a weight on a string) to mark an accurate plumb line. Hang the plumb bob from ceiling to floor so that it passes ¼ inch to the side of the seam mark. Mark the top and bottom points on the wall, then snap a chalk line between these two points. (You can use a chalk line that acts as plumb bob; see Glossary.)

Beginner wallpaper hangers may want to construct plumb lines, as described above, ¼ inch to the side of every seam mark. You can then see how well your plan works, adjust it if necessary, and use the plumb lines as you hang the paper. If you don't construct a plumb line at every seam, you will have to reestablish your plumb line at various points as you hang the paper.

Another layout decision to be made is exactly what point of the pattern on repeat-pattern papers should be aligned with the top of the wall. This is mainly a question of individual taste, but in general it looks better if the dominant part of the pattern is located near the top. If walls and ceilings are not perfectly plumb and level, such prominent placement of the main pattern element may tend to exaggerate these faults—it might be better to arrange the paper to draw attention from any structural deficiencies. Once you have decided on pattern placement, use a chalk line snapped horizontally just below ceiling level around the perimeter of the room to ensure the alignment of the pattern as you hang the paper. If the room has crown molding (molding around the corner of the walls and ceiling), use the bottom of it as a guide. Don't try to conform to the room's inconsistencies—it will invariably look better if the paper is hung correctly, making adjustments as you turn corners.

Hanging the Paper

The Paste for the Job   Choose the correct paste. Powdered *wheat paste* is proper for all papers except nonporous papers such as foil films and vinyl papers. Special *vinyl pastes* (often referred to as vinyl "adhesives") are best for such papers because moisture trapped under nonporous papers can cause the organic ingredients of wheat paste to develop mildew.

Vinyl pastes come premixed, and premixed pastes are far easier to use than the pastes you must mix yourself. Wheat pastes are normally supplied in powder form and must be mixed before use. The usual proportion is one part water to two parts powder, but mix the powder into the water, not the other way around. Stir it just enough to mix it well and make sure it's lump free. Let it stand for thirty minutes to an hour before using it.

Although many papers now come prepasted, it is still a good idea to guarantee complete adhesion by pasting even these papers with your own paste.

Cutting and Pasting   Unroll a length of paper and hold it against the wall. Adjust the paper up and down until the pattern aligns with the horizontal mark on the wall (or crown molding), as you planned. Mark this spot on the paper with a pencil and take the roll to your table or work surface. Allowing an extra 2 inches beyond your mark, cut off the top of the paper. Measure the height of the wall to be covered. Allow another 2 inches at the bottom of the paper and cut it to this length. This is your first strip; it should be 4 inches longer than the height of the wall to be covered. To make it easier to work with, roll this and subsequent strips up the opposite way to which they were supplied before proceeding with pasting and hanging. More experienced paperhangers measure and cut a few rolls at a time, then paste and hang them. Beginners should cut, paste, and hang one roll at a time.

Now lay this strip face down on the table, hold the top end securely (a small clamp can be used to keep the paper from rolling up on itself), and apply paste. Work from the center to the edges, covering slightly more than half the complete strip. Make sure you cover the entire surface as evenly as you can. Don't be too messy at the edges; keep a clean bucket of water and a damp sponge handy to remove any excess paste. Now fold the end of the pasted half back on itself so that the end of the paper lies at the midpoint of the full strip. Turning the strip end for end, repeat the process, folding the bottom end in to the middle as well.

This folding process, which results in the pasted surface facing itself, is known as *booking* (see Figure 39), and helps spread the paste evenly while preventing much evaporation during the 10 minutes or so that the paper is left to *cure*. Curing allows the paper to soften somewhat, which makes it easier to manipulate, and to swell to its full dimensions. (None of the vinyl papers needs to cure.) If the paper has a selvage to be trimmed off, this is the time to do it. Use a sharp razor blade in a razor knife against a straightedge, cutting through both folded-over layers at once.

If you're using prepasted paper and adding paste, cut and paste the strip just as you would unpasted paper. If you're using prepasted paper and *not* adding paste, cut the strip to length (with its 2-inch margins at each end), roll it up, bottom end first, and soak it in a water-box for the prescribed amount of time. That will vary from paper to paper and manufacturer to manufacturer; follow the instructions with the paper. The water-box may be any simple trough long enough to contain the width of paper you are using, then half filled with warm water. (Plastic ones are usually available from the paper dealer.)

Beginning to Hang   Starting with the unfolded or unrolled top end, position the pasted and cured strip against the wall, lining it up with your plumb line. Using your hands to slide it into position, brush any folds or wrinkles out by working from the center to the edges. The paste takes a while to set, so take your time to get it just right. You want to be sure that the strip is positioned perfectly, with regard to both the pattern and the plumb line. Use a slightly dampened sponge to wipe up any paste that may find its way to the surface of the paper. Slowly unfold or unroll the bottom half until the entire strip is brushed out, wrinkle-free and perfectly aligned on the wall. Do not trim the top and bottom edges until the adjacent strip has been hung, in case you need to reposition it.

Until you are very practiced at hanging, it's not a good idea to cut too many strips at once. What is a good idea, since it may save you paper with repeat-patterns, is to cut

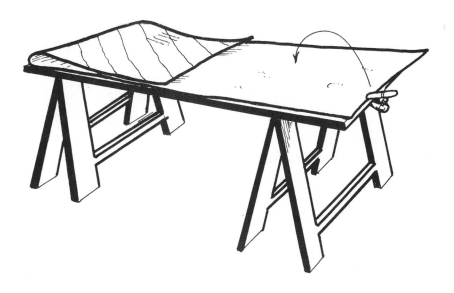

Figure 39   Booking

subsequent strips from *alternate* rolls. For the second strip, hold the first roll up to the wall and mark it for cutting just as you did for the first strip, but align the pattern carefully with the already pasted first strip. Mark it and cut it to length (including the extra 4 inches). Cut the third strip from a second roll, aligning its pattern with the proper edge of the second strip. If you alternate in this way, there will be less waste since one of the rolls will almost always produce an extra strip at the end.

Once you feel confident enough, it will save time to cut several strips at a time, laying them face down on the table in sequence, with the first one to be hung uppermost, ready to be pasted.

Seaming and Trimming The second strip is hung the same way as the first except that you must position it against the edge of the first strip. Doing this so that the edges just raise one another slightly will allow for any subsequent shrinkage. After 15 minutes, rub these edges flat with a seam roller. If a gap opens up as the paper dries you need to overlap the second edge a little more.

You can also finish the seams by double-cutting: Overlap the seams by ½ inch or more, then trim through both layers (in the middle of the overlap). Peel off the remaining top layer first. Then pull the remainder of the top layer back just a bit to remove the trimmed edge of the underlying layer. That produces a perfectly butted seam, but also creates a slight pattern mismatch that may not be acceptable unless it occurs in a corner or other obscure spot.

The easiest way to trim off the top and bottom edges (and at door and window moldings) is with a razor knife. To prevent tearing the paper, you should constantly change the blades. (One blade every sheet of paper is a good rule; some professionals change blades after every cut!) Run the blade against a wide-bladed spackle or joint knife as a guide. Another useful technique is to brush the paper down against the edge of the trim, creating a crease in the paper at the joint, then pull away the paper and use paper shears to trim at the creased line.

Dealing with Inside Corners It is often difficult to hang a complete width around an inside corner. For one thing, the paper may wrinkle. For another, when you round the corner, the next strip may no longer be completely vertical. To avoid these problems, cut and overlap at the corners (see Figure 40).

Before you hang the corner strip, measure the distance between the last strip and the corner, both at the top and at the bottom. Cut a strip of paper through its length so that it is a ½ inch wider than the wider of these two measurements. Hang it—it should come around the corner at least ½ inch.

Now you must establish a new plumb line around the corner. Measure the remaining piece at its shortest width; measure out from the corner and mark the wall. (The new mark may vary slightly from your guideline mark.) Snap a plumb line ¼ inch outside your measured mark.

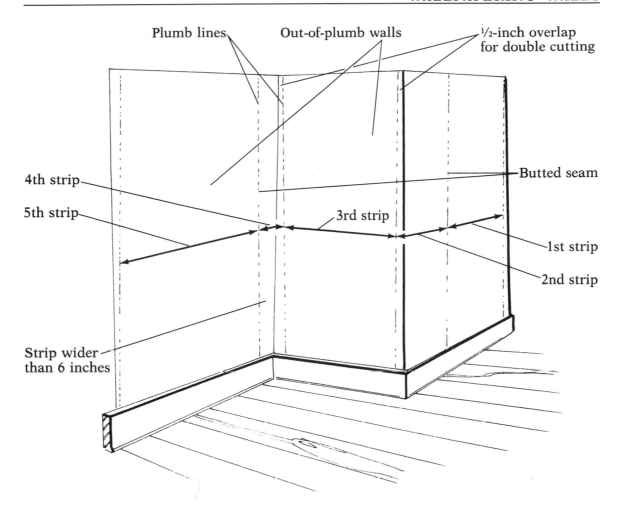

Figure 40   Locating plumb lines when working around corners

Now hang the piece you cut from the strip so that its uncut edge—on the wall around the corner—is perfectly vertical. It should line up ¼ inch inside the plumb line, and its cut edge should fall exactly at the corner. Trim it at the corner, double cutting as described above. If necessary, accommodate any slope in the wall.

Dealing with Outside Corners, Windows, and Doors   Outside corners have to be negotiated by cutting slits exactly at the corners in the excess paper at the top and bottom edges of the strips. Those slits will allow the excess 2-inch top and bottom to be creased over and then sheared off. This is also how windows and doors are dealt with as well; crease the paper to ascertain its edge, and then make any slits necessary to allow the paper to be turned, stopping the cuts at the creases (see Figure 41). If curved cutouts are required, the slits must be cut without a crease, then folded around the corner. A second strip is pasted over the slits to hide them.

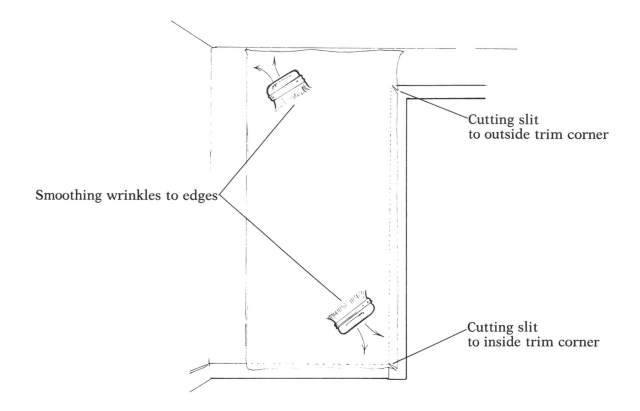

Smoothing wrinkles to edges

Cutting slit
to outside trim corner

Cutting slit
to inside trim corner

Figure 41   Smoothing and cutting

You must reestablish your plumb line, as described above, after working around an outside corner, a window, a door, or any other obstruction. At an outside corner, measure the width of the next strip out from the corner to establish your mark, as you would with an inside corner. For the space over a door or the spaces over and under a window, subtract ½ inch from the width of the next strip to allow an overlap for double cutting.

It's often convenient and less obtrusive to let mismatches fall at the corner over a door. Rather than fit two full-length strips around a door, you can fit one piece around the door, then measure and hang a narrow strip (butted against the seam of the first piece) so it protrudes ½ inch beyond the outside edge of the door trim. Measure the width of the next piece from the outside of the door trim. Establish a new plumb line, then hang the new piece and make a double-cut seam at the corner of the door.

Be very careful when papering around light fixtures, electrical outlets, and switch boxes, from which the cover plates have been removed. Since the paper is wet, make sure that the power is turned off while you paper and trim these areas.

138

REPAIRING WALLPAPER

Any wrinkles that occur during the actual hanging should be eliminated by removing the paper and repositioning it. If you damage any strip badly, it should be replaced. Any blisters that show up during hanging indicate that air is trapped under the paper. Use your smoothing brush or a broad-bladed knife to push them out the side of the paper before you hang the next strip.

If blisters remain after the whole wall has been papered, simply cut the blister with a very sharp blade—along a pattern line if possible. Inject some wallpaper paste or white glue into the cut and roller the blister down. You can repair small tears, and corners that have lifted away from the wall, in the same way. Try not to crease the paper; immediately wipe up any excess paste or glue with a clean damp rag or sponge.

There is a wide range of cleaners available for different kinds of papers. Not all papers are washable, though many may be made so by the application of a protective coating. Once again, read labels carefully to be sure that any coating you use is compatible with the paper.

13

PANELING

TYPES OF PANELING

The term *paneling* covers a variety of wood and imitation-wood products that are used to cover interior walls. Primitive wood houses usually had the same simple wood walls on the interior as on the exterior, but sophistication has led to other forms of interior wall finishes, including plaster and paper, and finally back to wood again. However, though today's wood paneling tries to resemble those early interior walls, it is often not as simple as it appears. The term paneling now most frequently suggests 4-by-8-foot sheets of various wood products (most commonly plywood), but it can also refer to solid wood used for interior walls.

Solid-Wood Paneling

The simplest form of wood paneling is composed of strips of *solid wood* nailed to the wall. The strips are usually in the form of planks or boards, somewhat less than 1 inch thick and varying in width from 3 inches to as much as 12 inches. Applied vertically, horizontally, diagonally, or even in a combination of directions, the wood can be of almost any species that can be sawn into plank form. Probably the two most commonly used varieties are pine and redwood.

Square-edged boards obtained directly from sawmills, or recycled from old barn siding, are occasionally used, but most of the paneling used today has some form of edge joint so the planks can form a continuous wall, showing no gaps (see Figure 42).

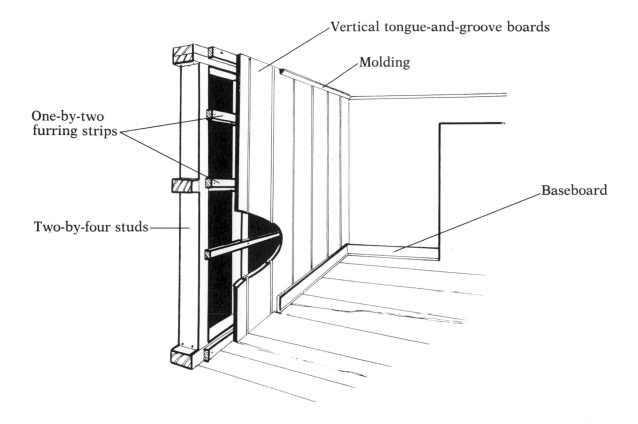

One-by-two
furring strips

Two-by-four studs

Vertical tongue-and-groove boards

Molding

Baseboard

Figure 42   Solid-wood paneling

The Expansion-Contraction Problem   In the normal course of events, all wood expands and contracts in response to changes in the moisture content of the surrounding air. Fitting the edges of boards together to hide this shrinking and expanding is not too difficult, but allowing the wood freedom to move and, at the same time, keeping it firmly fixed to a wall is a bigger problem. The narrower the board, the less it changes size—but more work is required to panel a room, since more boards are required. Simply increasing the width of the boards is not a good idea; if wide boards are firmly fixed to the wall, they may develop cracks if not allowed to expand and contract. In response to this problem, and also as a way of providing greater design potential, frame-and-panel wall coverings were developed.

Frame-and-Panel Systems

The essence of the *frame-and-panel system* is that narrow frames (subject to fewer dimensional changes than wide boards) hold wide panels that are free to expand and contract (see Figure 43). The panels are not fixed in any way, simply held in place in grooves cut in the edges of the framing. Thus, they can slide back and forth in response to changes in humidity.

With the frame-and-panel system a room can be paneled using very wide boards with no fear of unsightly cracks or gaps developing. At the same time, the arrangement of the frames and panels can be varied almost without limit, allowing for unusual design ideas. Over the centuries, panels have been painted and carved, contrasting woods used, and such grand effects achieved that frame-and-panel walls are generally considered one of the finest ways to decorate a room. Frame-and-panel wall coverings, however, require a high level of woodworking skill not generally available or even affordable today.

Sheet Paneling

In response to the need for a paneling that is inexpensive and easy to install, *sheet paneling* has been developed. Its success has been based on new manufacturing technology. The sheets of material are simply nailed or glued to the walls in large sections, requiring very little

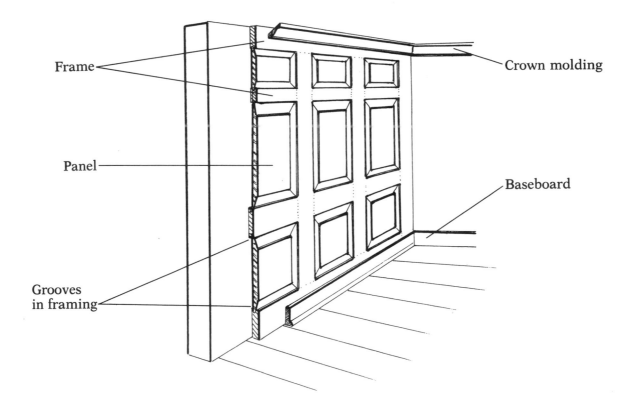

Figure 43   Frame-and-panel wall

joinery. The sheets can be made of plywood, hardboard, composition board, and even plastics.

These sheets provide not only surface covering and decoration but also additional functions such as fireproofing, soundproofing, and insulation. The sheets can be manufactured to mimic the simplest board paneling; other types use real wood-veneer surfaces; and in between are sheets with endless varieties of pattern, texture, and design.

Although other sizes are occasionally used, sheet paneling is usually made 4 feet wide by 8 feet high. The thickness of the sheet varies, depending on the material used; plywood paneling, for example, is most typically available from $3/16$ to $3/8$ inch thick. Sheets are nailed (or occasionally glued) either directly to the wall or to a system of narrow wood pieces called furring strips, which provide a framework for the sheets (see Figure 44). The joints where the sheets meet are, in some cases, designed to be relatively inconspicuous and are left as they are. In other cases, they are hidden with various types of molding strips.

Some sheet paneling, notably exotic veneers, can be extremely expensive. In the main, however, sheet paneling is comparatively low in cost. Low cost and the ease and speed of installation are its chief advantages.

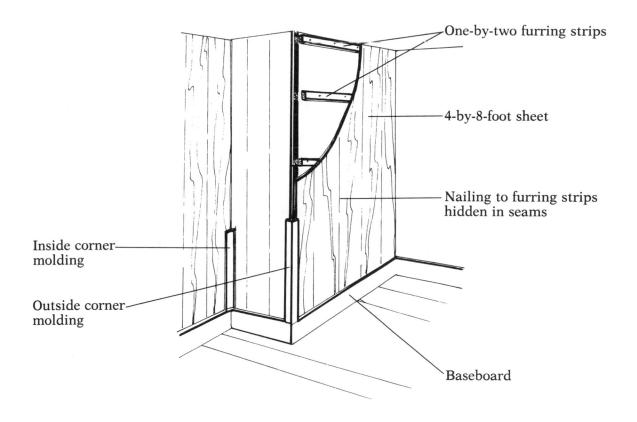

Figure 44   Sheet paneling

ASSESSING THE CONDITION

Solid-wood paneling is often painted, and a quick and easy way to improve its appearance is to repaint it. Not as simple, but often more dramatic, is stripping the paint and restoring the wood surface. Sometimes, however, the wood is so damaged or deteriorated that it's better to remove it and start again with new paneling or some other treatment. Sheet paneling, on the other hand, is not usually painted; when the paneling is in need of a face-lift, it's harder to renovate or repair. Replacement is often the best course.

Solid-Wood Paneling

Board paneling, once dirty or stained, can be refinished, but it is usually preferable to paint it. If it has already been painted and you are considering stripping it for a natural wood finish, try to find an unpainted area—such as behind a baseboard or in a closet—to be sure what the surface really looks like before you begin the onerous task of paint removal. Not all wood paneling was designed to be left natural. Matchboarding, for example, a precursor of plywood, was used in many turn-of-the-century houses. It consisted largely of thin oak strips joined vertically to look like even narrower strips. Much matchboarding was intended to be painted right from the start, and stripping may reveal unsatisfactory surfaces.

Similarly, frame-and-panel walls enjoy a long tradition of having been stained and painted. Colonial paneling was often very brightly decorated; much of the intended effect is lost if its decoration is removed. At the same time, there was also a tradition of more somber stained and polished paneling that would lose much of its richness if inappropriately painted.

If frame-and-panel walls have damaged panels or rotted framing, they'll need professional repair. The cost may be a wise investment if you wish to preserve the integrity of the house, since duplicating a frame-and-panel wall can be very expensive. Simpler solid-wood paneling, on the other hand, can be more easily repaired and should not necessarily be condemned out of hand. Of course, the loss will be less if you decide to remove it. Knotty-pine paneling is cheap to replace, but it may be hard to blend a replaced section with older areas. Weigh the cost of repair (and refinishing) carefully against the cost of total replacement.

Sheet Paneling

With sheet paneling the situation is different. Deterioration of such walls is often the result of poor installation. The sheets themselves may still be intact, but if the seams have opened up or the sheets are bowed or loose, there is little to do except remove them. You may be able to save the old material and start again. Its condition may not be as important as its effect on the decor. If it is out of fashion, there is little that can be done to change its appearance. Usually, it must be removed rather than covered up.

REPAIRS

Repairing Solid-Wood Paneling

Aside from simple repairs that might be part of the proper preparation for painting—such as filling small holes, reattaching loosened pieces of molding, and filling and smoothing various surfaces—almost any repairs needed by a frame-and-panel system will require the attention of a skilled woodworker. Other solid-wood paneling, however, is easier to repair. In fact, simple board paneling is more likely to need repair since its construction is not as sophisticated—and therefore not as durable—as frame-and-panel woodwork. A diligent handyperson can make the repair with basic woodworking tools.

Repairing Board Paneling

The most common fault is a board that has become unfastened from the wall. This is often the result of shrinkage. The board becomes so separated from its neighbors that it lacks sufficient restraint to hold it in place.

The quick cure is to face-nail it back into place. Be sure to nail through the board into something secure—the wall framing or the furring strips to which the paneling was originally attached. Use a nail set as you finish driving the face nails so you don't damage the surrounding wood with the head of the hammer. Set the nail below the surface, then fill the hole with wood filler of a matching color.

Face nailing will secure the board neatly but will not necessarily take care of the gaps on either side of the board occasioned by its shrinkage. When rough-sawn or other square-edge boards have been applied, it is usual to put a sheet of black building paper behind them, to minimize the appearance of any subsequent gaps. (This is unnecessary with tongue-and-groove or other fitted joints.) If there is no building paper and the paneling has shrunk substantially, nothing will improve the wall's appearance so much as removing the boards, installing black paper, and reattaching the boards, this time more closely together.

Replacing a board in a wall paneled with tongue-and-groove paneling is tricky but not extremely difficult. Remove the old board as cleanly as possible, without damaging the tongue or groove of the neighboring boards. Now cut off the wood piece that forms the back of the groove on the replacement board. Fit the tongue of the replacement into the groove of the neighboring board and drive it in, using a scrap piece of softwood to avoid damaging the edge of the replacement board. You should now be able to push the grooved side into place over the corresponding tongue of the neighbor on the other side (see Figure 45).

Repairing Sheet Paneling

In a wall with sheet paneling, it is often the molding that becomes damaged or lost. There are various moldings holding or finishing different parts of the system: inside and outside,

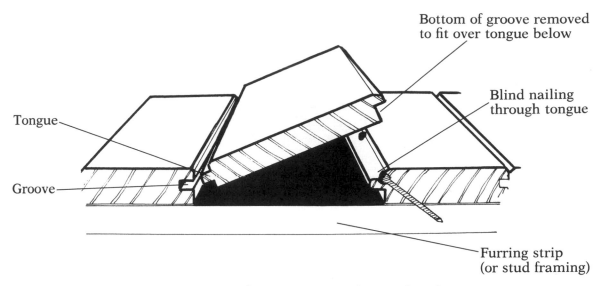

Figure 45   Replacing a tongue-and-groove board

corner moldings, crown cap, or base pieces, battens, and edging strips. Most of these are standard and are readily available at building suppliers and lumberyards. If you can't match a particular molding and the damage is serious, consider replacing all the moldings (see chapter 16). The cost of new molding is not so great.

Individually damaged sheets are harder to repair. If a replacement panel is expensive (or not available), it may be possible to remove the sheet and repair it from behind by gluing on a thin "skin" of some additional material—like thin plywood—if this can be done without altering the plane of the wall. Occasionally, panels will partially delaminate; you may be able to inject glue between the laminations and stick them back together again.

Much sheet paneling is supplied prefinished. Should this finish ever deteriorate, the work entailed in stripping and refinishing will probably be much greater than replacing the paneling.

INSTALLING NEW PANELING

Preparing the Existing Wall

It requires neither a great deal of skill nor any special tools to install either solid-wood board paneling or sheet paneling, but an assistant can be very helpful maneuvering and securing sheet paneling. For the inexperienced, the jobs do require time and, as always, care. If you're installing solid-wood board paneling horizontally or diagonally on a framed wall, no preparation is necessary, assuming the wall has a level surface. Every board can be nailed to a vertical framing member in the wall. If you're installing such paneling over a masonry wall, or vertically over any kind of wall, you'll need to provide furring strips so that there is something to which to attach the paneling.

Sheet paneling will always require a subsystem of furring strips. If it is very thin, sheet paneling may also require the additional support of a layer of gypsum between it and the furring strips.

Before Applying Furring Strips   If you're paneling a masonry wall, and there is the slightest hint of moisture, you should install a moisture barrier—a continuous sheet of plastic under the furring strips. If the moisture is excessive, treat the cause and perhaps seal the wall with a masonry sealer. Even then, the sheet of plastic would be extra insurance.

Before going ahead with the furring, make sure that the wall is sound and that any electrical wiring in the wall is well protected. Reposition outlets and switches, if necessary, so they will protrude as needed through the new paneling. This may entail relocating existing boxes or attaching extensions, a job best left to those with knowledge and experience. Be aware that in many localities a licensed electrician is required to perform any work on residential wiring or rewiring. Finally, remove baseboard and other trim, as necessary.

Since furring also provides a means of leveling an uneven wall, look over your walls carefully. Then, as you attach the furring, you can shim it out with small wedges where necessary to obtain or maintain a level surface. Use a long piece of straight lumber or a length of taut string to check that the strips are flush with each other.

Installing Furring Strips   Furring can be made out of any sound wood 1 inch thick by 2 inches or 3 inches wide. The width is not important, but it must all be of equal thickness to provide a level surface for the paneling. When installing solid-wood paneling, only horizontal strips need be used. When installing sheet paneling, both horizontal and vertical strips should be used.

Horizontal strips should be nailed securely to the studs using 8-penny nails (or masonry nails, if it's masonry). Place one strip at the top of the wall, one at the bottom, and at every 16 inches in between. The vertical strips should also be inserted flush with the horizontal strips every 16 inches. They must be spaced, however, so that a strip will fall at every point where two sheets of paneling meet. To do this, you'll have to plan your layout in advance.

Plan the layout of the sheets so that the partial widths that may be necessary are not too narrow. Start by measuring from the center of each wall, in 4-foot increments, to the nearest corner. If, after the last full width has been measured, there remains more than 2 feet to the corner, locate the panels on that wall so that a seam is at the very center of the wall. If the remaining distance is less than 2 feet, center a sheet on the wall.

Doors, windows, and other openings must also be considered when determining where panels will be located. It usually looks better, for example, if sheets are so arranged that joints occur in the middle of openings rather than at one side, even though it may seem easier to align a sheet against the edge of a door. After you've planned the layout, pick specific sheets for each location. Usually, not all the sheets are identical—some will be best next to each other, others will be good for the narrow areas.

Once you've decided exactly where the sheets will meet one another, install vertical pieces of furring strip between the horizontal furring at these points. Should this not be possible— as when, for example, you are paneling a framed but otherwise uncovered wall—then the location of the studs will also figure in your determination of sheet positioning. Normally, however, there will be at least a layer of gypsum board to which vertical furring strips can be glued. These strips do not have to be solidly backed, since sufficient support will be provided by the horizontal furring to which the sheets are nailed. But you should always provide some form of base under the seams. Some sheet paneling is made with jointed edges, which helps keep them tightly aligned, but often the edges are plain. Nothing looks worse than sheet paneling whose individual sheets are not flush.

Attaching Solid-Wood Paneling

Ideally, each strip, especially if being installed vertically, should consist of a single, full-height board. Most wood designed as wall paneling is available in sufficient lengths to make this possible. Some species, such as redwood, may be available in extremely long lengths, making multi-storied seamless paneling possible.

Square-edge paneling, such as recycled barn siding, is useful for certain primitive effects. Whenever possible, however, it is best to use boards with fitted joints. These make a stronger, more integral wall covering. And if excessive shrinking occurs, gaps are less likely to appear between the boards. Knotty pine made specifically for paneling often has a tongue cut on one edge and a groove on the other, so adjacent pieces will fit into one another. Furthermore, to minimize the visibility of any potential gap, the edges are beveled so that when they are joined they form a V-joint, creating a gap-obscuring shadow. Other edge treatments take this principle even further and sport more ornate moldings that make the detection of any gap almost impossible.

Start paneling a wall by placing the groove side of one board against the left corner, then work away from it so the tongue always shows on the last board attached. This way nailing can be done at an angle through the tongue (see Figure 44), thus avoiding any nails on the surface of the wood except in the corners where the first and last boards are necessarily face-nailed.

If the paneling doesn't reach to the ceiling, where it will be covered with molding, a temporary cap strip should be installed before you start so that every board can be tightly aligned and butted to it, leaving any discrepancies at the bottom of the paneling, where they will be covered by baseboard (see Figure 46). Snap a horizontal chalk line around the room's perimeter at the height of the top of the first course of boards. Check the line's level, then nail a one-by-two or one-by-three with duplex (double-headed) nails so that its bottom edge aligns with the chalk line. Remove the cap before you install the top course of boards. Butt the top boards against the bottom boards, leaving any irregularities at the top to be covered by crown molding.

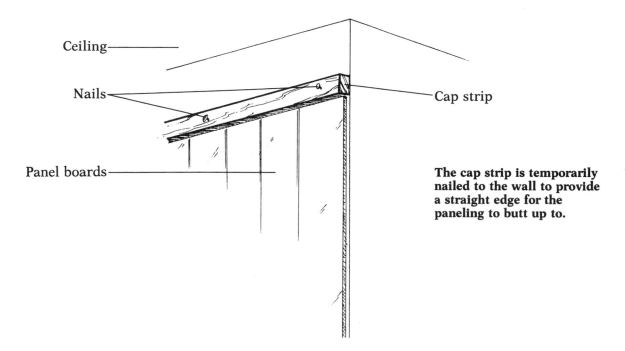

Ceiling

Nails

Panel boards

Cap strip

The cap strip is temporarily nailed to the wall to provide a straight edge for the paneling to butt up to.

Figure 46 Temporary cap strip

Attaching Sheet Paneling

With sheet paneling, especially the thinner types, it is advisable to store the sheets for a couple of days in the room in which they will be installed. If stored completely flat, with spacers placed between each sheet, the sheets will be conditioned to their environment and less likely to swell, shrink, or otherwise deform after being put up. Generally, you'll be working from the center of each wall toward the corners.

Adhesive or Nails   The lighter weight panels can be installed with adhesive alone, but heavier sheets should be nailed. Nails colored to match the sheet are usually supplied with the paneling. Even if you feel adhesive will be sufficient, there's no reason not to nail the sheet at the top and the bottom, where it will be covered by molding anyway. The face veneer of plywood paneling is usually arranged so that even when you have to nail through it, you can do so in shallow seams on the face of the panel. This keeps the nails as inconspicuous as possible. It is always good practice to nail at regular intervals and, where possible, at places that will be covered by molding. Leave a gap between the top and bottom of the panel and the ceiling and floor, but one no greater than what will be covered by the molding (see chapter 16).

Cutting Sheets   When you have to cut a sheet, try to produce as smooth an edge as possible. If you're using a handsaw, saw from the face side; if you're using a power saw, saw from the back side. Otherwise you'll produce ragged edges on the face because a handsaw cuts down and a power saw cuts up.

To look its best, all paneling should be installed perfectly vertical, using a plumb line or a spirit level. Since not all walls are square or plumb, however, you will undoubtedly have to cut certain edges to fit. A sloping but straight edge should present little problem: Take two measurements to the corner, one from the top of the last sheet installed, and one from the bottom edge. Transfer these measurements to the new sheet, connect the points with a snapped chalk line, and make the sloping cut.

Scribing   An irregular edge of almost any kind can be matched by the process known as *scribing* (see Figure 47). Hold the panel to be scribed against the irregular edge but keep it perfectly vertical. Part of the panel's side will no longer be in contact with the irregular edge. Open the legs of a pencil compass (or similar, but more professional, *dividers*) so that when they are held perpendicular to the edge of the panel, they span the largest gap between the panel and irregular edge. Holding the compass or dividers perpendicular, run a pencil line down the entire panel, letting the compass move in and out as the irregular edge requires. If you now cut exactly to this line, the panel should match the irregular edge.

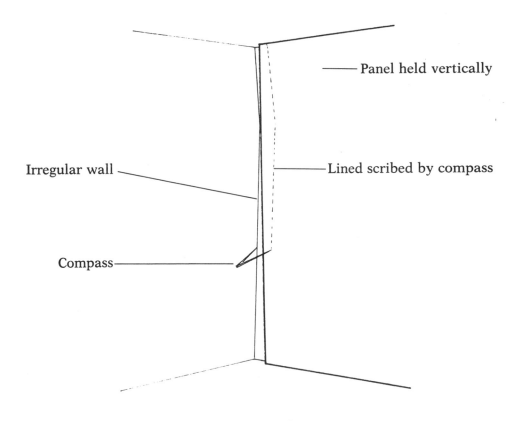

Figure 47   Scribing

14

TILING

Much of the information given in chapters 3 and 4 is relevant to the tiling of walls, and you should refer to those chapters where necessary.

TYPES OF TILES

Tiles suitable for use as wall coverings include many of the ceramic tiles described in chapter 4 as well as varieties closely related to types mentioned in chapter 3.

Ceramic wall tiles are made in a great range of colors, patterns, and sizes, from tiny *mosaic tiles,* less than 1 inch square, to tiles in large square "units" (referred to as *unitized tile*) that are actually several tiles held together by some type of backing (see Figure 48). Most lines of ceramic tile also include tiles for inside and outside corners, as well as a complete range of special tiles to fit around bathtubs and other bathroom fixtures (see Figure 49).

Another type of tile is molded plastic tile. It is usually much less expensive than ceramic tile. Made in sizes similar to ceramic tile, plastic tile is applied in a bed of water-resistant mastic over waterproof board. However, most molded plastic tile is *not* suitable for areas subject to much moisture. Its appeal lies in the fact that it can be obtained in patterns and designs resembling almost any kind of wall surface imaginable, from brick to barn siding. The tiles tend to be lightweight and large enough to make the job proceed quickly, even if fairly large areas are being covered, and they very often come with an adhesive backing—all of which make application very easy.

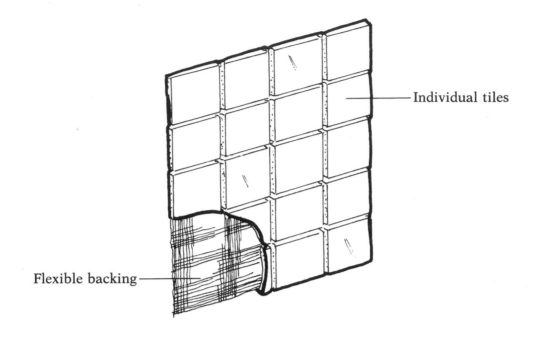

Individual tiles

Flexible backing

Figure 48   Unitized tile

Several other types of wall covering that are available in tile form should be mentioned, although they are not tiles in the strictest sense. One is brick tile, made of mineral composition, and to all intents and purposes the same as real brick, except that it is generally only ½ inch thick. Others include cork tile, usually supplied in 12-inch squares to be glued to almost any wall surface (as a bulletin board, to improve acoustic insulation, or simply for looks), and metal tiles that have a baked enamel finish.

MAINTENANCE AND REPAIRS

The procedures for assessing the condition of and making common repairs to ceramic tile are described fully in chapter 4.

Damage to the various forms of molded plastic wall tiles can frequently be repaired in the same way as resilient floor tiles, described in chapter 3. The key to these repairs lies in keeping extra tiles to use as replacement material and knowing the correct adhesive to use for the particular application. Damage to a small area is often easily repaired, and if the surrounding area is in good condition, it usually makes sense to do so. But if the whole wall has become shabby or has been widely damaged, complete replacement of the tile is often the best course. When applying new tile, you must start with a clean wall, removing old adhesive, wallpaper, loose paint, and any substances that may react with the adhesive

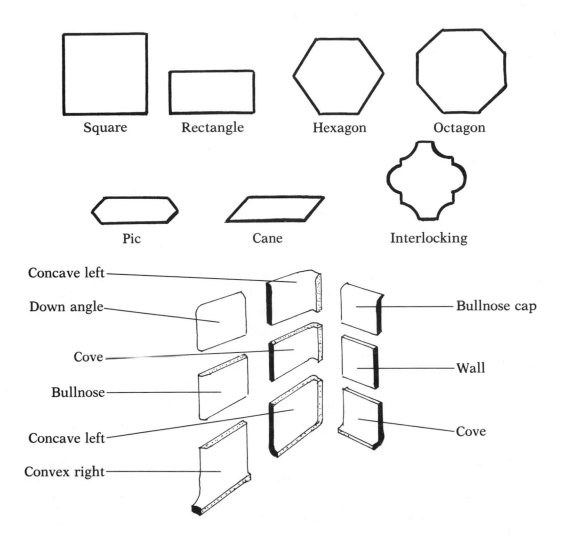

Square   Rectangle   Hexagon   Octagon

Pic   Cane   Interlocking

Concave left
Down angle
Cove
Bullnose
Concave left
Convex right

Bullnose cap
Wall
Cove

Figure 49   Common tile patterns and shapes

you intend to use. (You should always check the compatibility of the adhesive with the surface on which it is to be used; read labels carefully.)

INSTALLATION OF NEW CERAMIC WALL TILE

Preparing the Wall

Setting tile requires some skill and, for the less experienced installer, patience and care. Pay particular attention to layout (described below). You'll need to have a special toothed trowel to spread adhesive, and you should rent a tile cutter.

Although ceramic tile applied to walls doesn't have to withstand the same rigorous use as floor tile, there are nevertheless certain necessities for a successful installation. Of these, the first and most important is the condition of the wall.

Judging the Condition of the Wall   How fit the wall needs to be depends not only on the kind of tile you are using but also on the location of the wall and its intended use. For example, tiling over a wall that is less than perfectly even is fine if the tiles you're using are those composed of several tiles set in a rubber-backed unit, since the backing will absorb much of the wall's unevenness. Another example: If you're tiling a bathroom, you must be sure you can waterproof the surface perfectly, which may require special preparation.

Good Bases for Tile   Gypsum board in good condition is an ideal backing for most tile. However, remove any previous layers of wallpaper and clean painted surfaces thoroughly. Shiny paint surfaces should be deglossed. Trisodium phosphate (available as TSP in most hardware stores) is extremely effective for this job, but follow all precautions on the TSP packaging: Wear an old hat, safety goggles, long sleeves, and rubber gloves; cross ventilate the room with a window fan set to exhaust. Always follow its use with a thorough washing.

If a wall is in need of too much repair or cleaning, it may be simpler to install a new layer of gypsum board (see pages 108–14). Where conditions of extreme moisture can be expected, use of special water-resistant gypsum board is advisable. For an extra-strong wall, plywood can be used as the tile base. Tiles can be attached to plywood providing you first seal the surface. The best way is to apply a *skim coat* (a very thin, smooth layer) of whatever adhesive you plan to use for the tiles. Use a square smoothing trowel to lay on the skim coat. Never use asbestos board: it is a known carcinogenic.

New tiles can actually be applied directly over old tiles or over concrete walls, provided the surface is smooth and sound. But for walls in uneven or bad condition, a bed of mortar applied over wire lath may be a necessary foundation for any new tiling—a difficult technique best left to the very handy or to professionals.

Laying Out the Tile

The ideal wall would accommodate complete rows of tiles both vertically and horizontally, with no interruptions that were not increments of complete tiles. But in real life, no such wall exists—you'll almost certainly have to cut tiles at the ends of rows.

When laying out a wall, plan your tiling so the central focal area is done with complete rows of tiles. Any irregularity should be accommodated equally on opposite sides, and the size of those tiles at the sides should never be less than half-width. If a trial layout leaves you with less than a half tile at each side, move the row over to center a full tile at the middle of the wall rather than centering a joint between two tiles. When making trial layouts, be sure to allow for the width of the joints between tiles.

Often, the size of the wall to be tiled, or various obstructions and openings, will make complete rows impossible. In readjusting your proposed layout, bear the following principle in mind: The area on which the eye focuses first, as you enter a given space, should be the area that is tiled with the greatest care and regularity. It is usually preferable, therefore, to let any incomplete tiles fall at the base of the wall, since they will be less noticeable there than at the top. On the other hand, you may find places even better suited for hiding incomplete tiles, such as next to or behind fixtures, in corners, or behind alcoves.

Marking Out the Wall   After you've decided where the tiles should fall, establish a perfectly level horizontal line on the wall to mark the bottom of the lowest complete row of tiles. Mark the line to show the exact number of tiles plus joints. (If the bottom of the wall is level and you're tiling down to that point, you can treat it as your line.) At each end of the horizontal line, at the point where the last full tile ends, construct vertical lines marked to show an exact number of tiles (plus joints). (See Figure 50.)

If you've done your layout properly, the borders will be equal on both sides, and no less than the width of half a tile. These lines will help you tile the area with straight rows, both vertical and horizontal. Construct another vertical line at the center tile mark of the bottom line.

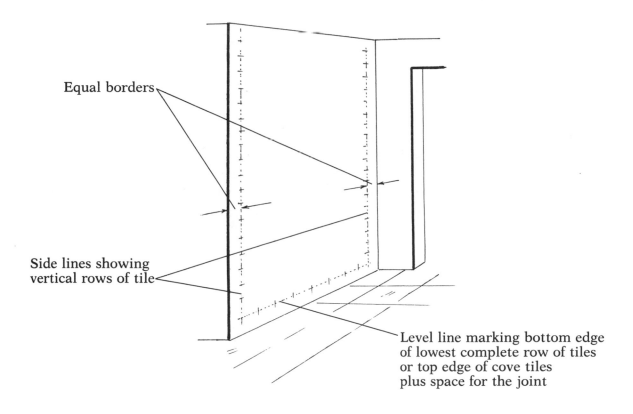

Equal borders

Side lines showing vertical rows of tile

Level line marking bottom edge of lowest complete row of tiles or top edge of cove tiles plus space for the joint

Figure 50   Tile layout

There's one exception to this procedure: If the bottom row of tiles are to be special *coved tiles* (see Figure 49), this row must be left full width. (Coved tiles are very difficult to cut.) Instead, accommodate any extra space either by adjusting the height of the tiled area, if you're not tiling all the way to the ceiling, or by trimming the top row, or the row just above the coved tiles. This same principle holds true for other specially shaped edge or corner tiles—measure for these first, then adjust the tiles within the borders of the coved tiles to fit as well as possible.

Tiling the Wall

Before tiling, make sure that you have the correct cement or adhesive (see description of these materials on pages 44–46) as well as the appropriate cleaning materials, and that the area you are about to tile has been fully marked out.

Setting the First Tiles   Before applying adhesive, cross ventilate the room with a window fan set to exhaust. Many adhesives call for a short waiting period while they are *setting up* (reaching the proper adhesive point). This may vary from product to product, so read the label instructions carefully. The best method is to apply the tiles in a pyramid pattern. Trowel on a layer of adhesive with the correct edge of a *toothed trowel* (a rectangular trowel, usually with teeth of two different widths). Allow the correct setting time, then place the first three tiles in the center of the bottom row—as shown by the drawn line whose execution was explained above. In order to assure complete contact between the bottom of the tile and the adhesive, use a slight twisting motion, but do not overdo this. If you slide the tile too far in any one direction, an excess of adhesive will be pushed ahead of it, preventing the proper joint being formed between it and its neighbor.

When you lay these first three tiles, take care to preserve the correct joint between them. With some tiles, this is guaranteed by small tabs at the sides of the tiles that establish the width of the joint. Should these be absent, use 6-penny nails gently inserted point first between tiles. Remove them when the tiles have set a little.

Building the Pyramid   Now lay the next tile over the middle tile, thus forming a pyramid. Now proceed to build up the pyramid on one side only until you reach the border tiles. Using this system, you'll always have two existing edges against which the next tile can be aligned. You can cover the other half of the pyramid with adhesive and allow it to set while you tile the first half. This system will work just as well with unitized blocks of several tiles as it does with individual tiles.

Don't allow the adhesive to set longer than it should. Don't apply more adhesive than can be tiled within the time limits given for the adhesive's usefulness. Note particularly that if you leave the cutting and fitting of the border tiles until the entire central area has been tiled, you should scrape up any extra adhesive in the border area to prevent it from hardening. It could cause difficulty when you come to fill in the borders. Rent a tile cutter if you have

1. Score along line to be broken

2. Place tile over narrow rod

3. Exert pressure on both
sides of scored line

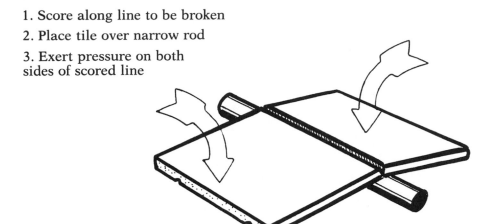

Figure 51   Method of cutting tile

a row or more of tiles to cut. If you don't have many tiles to cut, Figure 51 shows you a way to cut them.

Grouting   When all the tiles have been laid, clean off any excess adhesive with the proper solvent, and allow them to set for the recommended period of time. Now they must be grouted. This is the process of filling in the joints between all the tiles. Certain tiles, supplied in unitized groups of several tiles, have their own special filler strips that must be applied to the edges as they are laid. These tiles require no subsequent grouting, but all others do. For a full discussion of various types of grouts and the proper application procedures, see pages 48–49.

15

MASONRY

The role of masonry as a finished interior wall is rather limited. However, many basement walls made of concrete block, poured concrete, or sometimes stone or brick are left "unfinished." Perhaps more typical of a masonry finish wall is a brick wall that has been cleaned of its plaster covering. There are also masonry walls, originally left raw, that now find themselves, perhaps as the result of conversion from industrial to residential use, called upon to be more presentable. Finally, there is the use of stone or brick for localized effect as an interior finish, such as fireplace surrounds or kitchen work walls. Any of these may need attention, repair, or even improvement at some time.

REPAIR

Water Problems

Perhaps the most urgent repair that you're likely to face in residential masonry is one caused by water. It can range from mildew-causing dampness to a full-fledged leak.

Moisture can be a problem in the house wherever there is insufficient ventilation. Sources of moisture like bathrooms and kitchens can obviously create problems. When a heating system operates on a cold day, it can cause condensation inside a building with insufficient ventilation and no moisture barriers. Faulty plumbing systems are also a source of unwanted water. Beyond all these, there may be external drainage situations that can cause water damage to the house, especially if the house is built on a poured concrete slab with a full foundation below ground.

No matter what the souurce of the water problem, the cause must be dealt with before turning to the effect. Chronic moisture, seepage, or cracking problems will require evaluation by an engineer or contractor. But even an unskilled person can attempt most of the following repairs. You need little in the way of special tools except a cold chisel, trowel, and mixing bucket.

Moisture or Seepage   Of all the problems, the most difficult to deal with is poor external drainage. Usually the only permanent cure involves excavating around the outside of a building's foundations to effect repairs, a job that requires evaluation and execution by a qualified builder.

If the problem is limited to moisture or dampness, you have to decide whether this is caused by insufficient ventilation or seepage through porous masonry. To check for seepage, tape a large square of plastic to the masonry and leave it sealed there for several days. If at the end of this time the underneath of the plastic is dry, the damp condition is probably being caused by internal problems. Creating an air flow by opening area windows, installing a fan, and perhaps adding some temporary heat might well dry the place out if insufficient ventilation is the cause.

If moisture formed under the plastic, however, seepage through the masonry is indicated. Although painting with a waterproof *concrete sealant paint* may appear to work, the longevity of such a cure cannot be guaranteed. A better solution would be a coat of *patching mortar,* with a waterproofing additive such as latex or silicone added to it, troweled over the wall about ¼ inch thick. This job requires skill, and is probably better left to a professional. The coat is applied to a dampened surface, if the wall is not already damp, and treated further with a coat of concrete sealant paint as soon as the mortar has hardened— in about 20 minutes—but before it has fully cured.

Seepage Through a Crack   Should the problem be more localized, such as a crack, a simple repair may be all that is necessary. If water is actively passing through the crack, you must first stem the flow. Use a hammer and a cold chisel to enlarge the crack sufficiently in one spot so you can insert a drain tube (a short piece of hose is ideal). Apply *hydraulic cement* (available at hardware stores) around the hose (see Figure 52). The hydraulic cement will dry on contact with water, and will limit the leak to the drain tube. As soon as the crack is sealed around the hose, and the cement has dried, remove the tube and stop up the remaining hole with a plug made of more hydraulic cement.

Once the leak has been stopped, the crack may be repaired more completely with patching mortar. Chisel away any loose or crumbling material and widen the crack, if necessary, to about 1-inch width. Dampen the surfaces of the crack, then trowel in the mortar.

Chronic cracking or seepage may indicate a drainage problem around the base of the

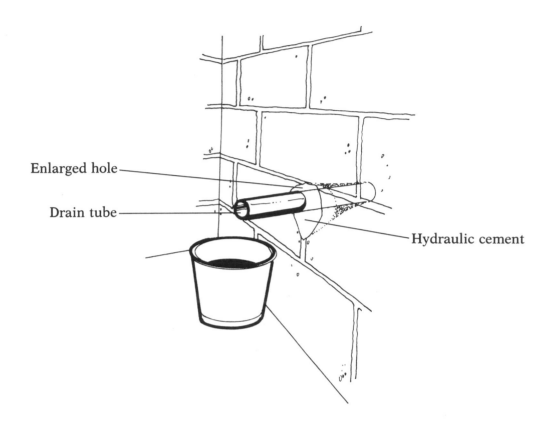

Enlarged hole

Drain tube

Hydraulic cement

Figure 52   Stemming an active leak

foundation or poor soil grading away from the foundation's top. A building engineer will have to evaluate the situation.

Repair of Other Cracks   An especially annoying kind of crack is the moving crack. This invariably penetrates an entire wall and may be caused by seasonal ground movement (the result of wet and dry periods or freeze and thaw), gradual settling, invasive tree roots, or even continual minor earthquake tremors. If a patched crack opens up again, it is obviously moving. If you suspect that it may be moving but want to be sure, simply mark both sides, measure the distance between marks, and then remeasure at regular intervals.

To repair a moving crack calls for patching with *asphalt sealer* applied over a *fiberglass patch:* Clean the entire area well, using detergent, and coat it with a thin bed of asphalt. Then cover the crack completely with a patch of fiberglass cloth (sold specifically for patching masonry). Cover the entire area with another thin coat of asphalt.

Very narrow cracks may be covered just with *mastic joint sealer*, heated with a *propane torch* and pushed into the crack with the end of a trowel. (When you use a propane torch, keep a fire extinguisher or bucket of water nearby.) When this is done, go over the repair with patching mortar.

Very large cracks in concrete block walls should be repaired the same way, but any exposed cavity in the block should be filled with *expansion-joint asphalt* (from a building supply store). Cut it into strips that can be inserted where needed, enlarging the crack a little if necessary to insert the material. Now close the crack with mastic joint sealer, as described above, and finish the job with patching mortar.

Repairing Brick

Brick deteriorates primarily when exposed to the weather, so it is unlikely that any interior brick walls will be seriously damaged. However, if a plaster covering has been removed to expose the brick, you may see various defects you'd like to repair. Certain bricks may have been removed, for example, and replaced with wooden wedges or blocks, called *noggins*, which formed anchoring points for woodwork that no longer exists. Then again, since the original intention was to plaster the wall, occasional defective bricks may have been used, adequate when covered with plaster but unsightly when exposed.

Choosing and Cutting Bricks   A word about bricks. There are many types, interior and exterior, and they are made in many colors and sizes. For the kind of repair we're dealing with, all you need worry about is matching the color of the original bricks and getting the right size. Even size is not of paramount importance, since brick can be easily cut with a little practice. Wear protective glasses and carefully score a line where you want the cut, using any sharp object, such as an old awl or metal set. Then deepen the score mark with a cold chisel; finally, break the brick at this point by giving the waste side a sharp rap with a hammer. It will help if the brick is supported by a cushion of soft material, such as loose earth or a bed of sand, while it's being scored. If the break is not clean enough, obtrusive lumps and nibs can gently be chipped off individually.

Replacing Bricks   Replacing a damaged or missing brick or two is not hard. You'll need some special tools: a 2½-pound hammer, a cape chisel or a plugging chisel or both, a small trowel, a joint or striking tool, and a large, square board to mix the mortar on.

The cavity in which a new brick is to be installed must be completely clean and free of old mortar. Chip out any remnants of the old brick and the old mortar, using a 2½-pound hammer and a cold chisel. Any chisel small enough to get in the hole will work, but professionals use a whole range of specially shaped chisels. Of these, the *cape chisel,* which is narrow and pointed, and the *plugging chisel,* which is broader, are probably the most useful. Be sure to wear protective glasses when chipping out masonry or wire-brushing and blowing out dust particles.

Immerse the new brick in water for a few minutes before you begin.

Being able to fit the brick in properly is largely the result of having made the mortar to just the right consistency. For small repair jobs it is best to buy small bags of ready-mix mortar to which you need only add water. While separate ingredients are much cheaper, they are generally available only in large quantities. Exactly how much water should be added to the mix will depend on the humidity, but be careful and add water slowly, mixing thoroughly with a trowel. The consistency should be like soft butter, which can be readily molded into any shape yet will not slump back on itself.

Wet the interior of the cavity well so that the old porous brick does not absorb too much moisture from the new mortar. Coat the sides, top, and bottom of the cavity well with mortar, along with the back of the replacement brick, but not so thickly that the mortar will prevent the brick from being fully inserted. Slide the new brick into place without leaving any air pockets around it (see Figure 53).

Striking the Joint   After replacing any bricks, you have to strike the mortar joint—that is, finish it off to match the surrounding joints. Brush off any excess mortar from the face of

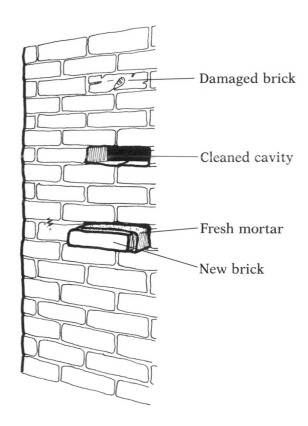

Figure 53   Replacement of damaged brick

the brick. Then draw a *brick jointer* (from a building supply store), or any similar hard smooth object no wider than the mortar joint, along the face of the joint, smoothing and firming the mortar into place (see Figure 54).

Pointing Joints   Exterior walls are struck in different ways, producing different profiles of mortar joints, but interior walls, built originally to be covered, are more likely to have a straightforward flush joint. If the mortar in the joints is generally crumbling or missing in quantity, you can point them by chipping out the old loose mortar, wetting the joint, and applying fresh mortar with a small trowel. Whenever you do any chipping of masonry, wear some form of eye protection. Use a narrow cold chisel to remove the old mortar to a depth of no more than ½ inch. A cape chisel is ideal for this job. Work carefully; you don't want to damage the bricks themselves. Use a stiff brush to remove loose particles from the surface of the remaining mortar before wetting the surface.

There are special tools called *tuck pointers,* made for pushing the mortar into the joint,

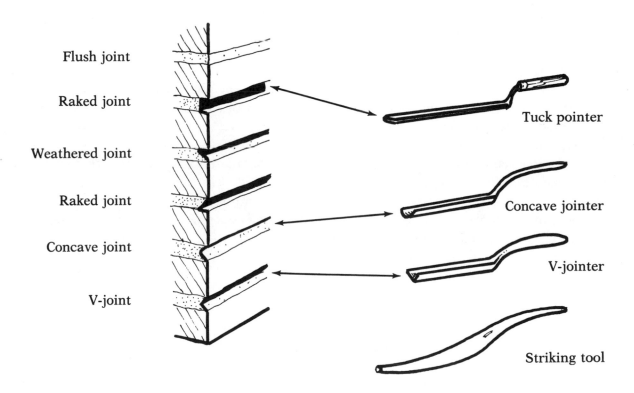

Figure 54   Striking tools and mortar joints

but any firm substitute will work as well, providing you avoid leaving any air pockets. Fill the vertical joints first, then fill the horizontal joints continuously across the entire width being pointed. Work neatly and try to avoid getting mortar on the face of the brick. Finally, clean the surface and strike the joints to match the existing joints.

TREATING MASONRY SURFACES

Although part of the charm of an exposed brick wall may be its variegated surface, you may still want to remove large unsightly stains and clean up begrimed areas. The process of cleaning almost all kinds of masonry—brick, concrete, concrete block, marble, slate, or tile—is essentially the same.

Start by scrubbing with household detergent. It is surprising what a stout brush and plenty of elbow grease can accomplish. If this doesn't do the job, there are numerous chemical cleaners formulated especially for masonry. All should be used with extreme caution, but avoid using muriatic acid, one of the most powerful of acid cleaners. Use the chemical cleaners only with adequate ventilation and plenty of protection (rubber gloves and a face mask). Follow the instructions on the container carefully. Afterward, wash the cleaned surface down thoroughly to remove all traces of cleaner.

Less toxic, but still requiring care in their use, are various poulticelike cleaning treatments that are applied to the surface to leach out the stain or dirt. After these have been allowed to work for a while, all you need do, generally, is to clean the poultice off. However, these various cleaning processes may not only clean off the dirt or stain but may also bleach out the brick's color. An effective way of resurfacing brick is to rub it with another brick of similar color and texture.

Highly polished masonry surfaces, such as marble, may need occasional repolishing with appropriate polishing compounds. Buffing compounds like tin oxide cream or jewelers' rouge can be used with a sheepskin bonnet fitted over a polishing-disk attachment in an electric drill. For more porous surfaces, such as brick and concrete, silicone-based masonry sealants can be brushed on like paint. Although the sealants will provide protection against future soiling and will reduce any tendency to produce dust or powder, they will impart a certain sheen to the surface, which may not be desirable.

16

TRIM

TYPES OF TRIM

The term *molding* refers specifically to any ornamental cross section (usually of wood, although plastic is also sometimes used) whose function is chiefly decorative. Depending on the style of the house, moldings can be found at many places—dressing up staircases, cornices, paneling, picture rails, and almost any other surface. Moldings are also used to cover joints wherever major elements, such as a floor and a wall, meet or wherever other elements, such as windows and doors, are inserted in walls.

When molding is used to cover joints, it is known as *trim*. The commonest forms of trim are *baseboards,* found at the junction of the walls and floors; *door trim* (technically known as *architraves*), the casing around the edges of door frames; and *window trim,* which conceals the edges of the openings cut in the walls to contain windows.

All three types—baseboard, door trim, and window trim—are made in an astonishing variety of designs, from extremely ornate to severely simple. But all three, regardless of design, play a very important function. Trim moldings are perhaps the most easily damaged parts of a house, suffering almost daily encounters with vacuum cleaners, moving furniture, and the passage of sometimes careless occupants. And the moldings must invariably be removed and replaced when any of the other work described in this book is undertaken.

Despite their occasionally complex appearance, the basic structure of moldings is quite straightforward. Once the structure is understood, repairs and replacement are easy, requiring only the minimum of tools used with care. You'll need a sharp chisel, a miter box, a fine-toothed saw (preferably a back saw), a nail set, and perhaps a lightweight finish hammer.

If you have to remove trim in an older house, be especially careful so you can reuse as much as possible. The only way to obtain an exact replacement may be to have it custom made, and this could be expensive.

WINDOW TRIM

Window trim is perhaps the most complicated of the three main types because of the many different window constructions that exist. However complicated the actual window, the function of the interior trim is always the same: to conceal the joint between the actual edge of the wall and the window's framework and give a finish to the window. The exterior trim holds the window and its framework securely in the wall opening made for the window.

Parts of the Window Trim

In its simplest form, window trim may consist merely of four strips of wood that encircle the window, each strip being nailed both to the edge of the window frame and to the surrounding wall (see Figure 55). Note that *window frame* refers to the *casework*—the outer frame that contains the *sash*. (The sash, the framework actually holding the glass, is often incorrectly referred to as the window frame.)

Beyond this simplest form, window trim can include an interior windowsill, more correctly called the *stool* (the *windowsill* proper is the exterior bottom part of the window frame), and, supporting the stool from below, an *apron*. Stools can be square-edged or rounded over; window trim and aprons may be plain or decorated with a variety of applied moldings.

Repair and Replacement

Dents and scratches may be repaired as you would in any other woodwork—filling, sanding, and repainting (or restaining). Loose pieces can simply be renailed. If necessary, the whole system can be removed and replaced, but the pieces must come off in the reverse order in which they were installed.

Here's how the window is trimmed: After the window frame is placed in its opening, it is temporarily held secure by wedges inserted between it and the wall's structure. The exterior trim is secured first. Then, on the inside, the stool is installed first, cut so that it overlaps the sill and fills the width between the sides of the frame. At the sides, the ends of the stool (called the *horns*) are cut to extend out along the face of the wall just far enough to provide a base for the bottoms of the side trim (see Figure 55). The side pieces go on next; they

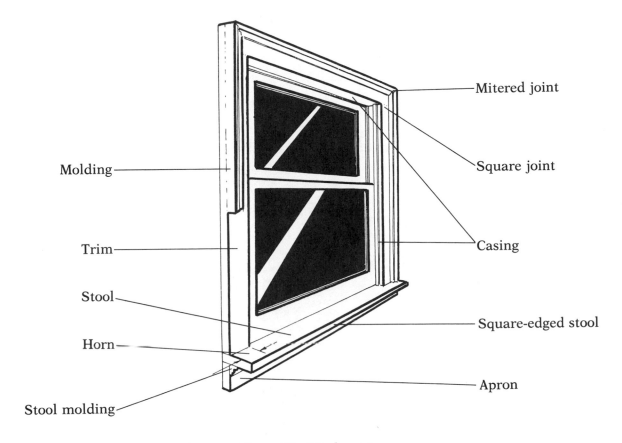

Molding

Trim

Stool

Horn

Stool molding

Mitered joint

Square joint

Casing

Square-edged stool

Apron

Figure 55   Window trim

rest on the stool's horns and extend to the top of the window frame, where they are joined by a horizontal piece that covers the top edge of the window frame.

If the trim consists of a single piece of wood, plain or shaped, the top corners are usually mitered. More complicated trim with applied molding, as shown in the illustration, is often left with square edges (less prone to produce an open joint) and only the applied molding is mitered. The important thing to remember when disassembling or repairing this kind of trim is that the applied molding must be removed first, separately from the trim; then the trim may be pried loose from the wall and from the edge of the window casing.

How carefully you remove the trim will determine whether or not it can be reused. Start by cutting the paint that covers the joint between the applied molding and the trim with a single-edged razor blade. Insert a fine but broad chisel between the pieces and pry the molding off, using something flat—like a broad-bladed spackling knife—under the chisel to prevent damage. An even safer method is to use a very thin nail set or metal punch—even an 8-penny common nail (file the tip down with a metal file)—to drive the nails holding the pieces so deeply into the wood that the pieces are no longer held and fall free. This

167

obviates any potentially damaging prying. The techniques for replacing window trim are similar to those for replacing door trim and baseboard, which are described below.

DOOR TRIM

Door trim can similarly be either extremely simple or very elaborate. Apart from the substitution of the door sill, or threshold, for the windowsill, stool, and apron, the basic construction and function of door trim is identical to that of window trim (see Figure 56). But at a doorway several other elements often have to be accommodated, such as baseboards, transoms, or fanlights.

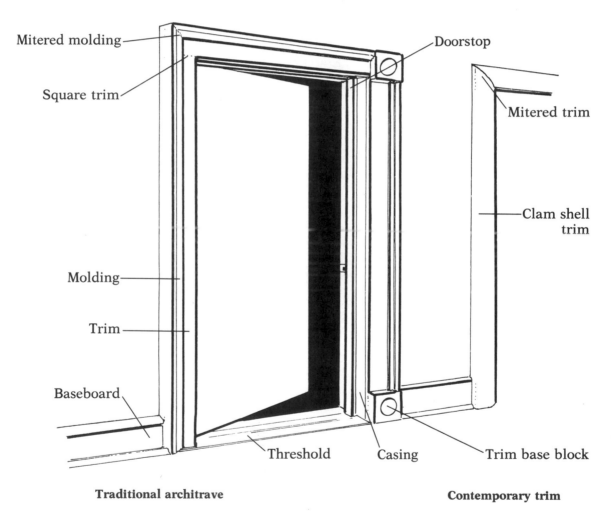

Traditional architrave   **Contemporary trim**

Figure 56 Door trim

Parts of the Door Trim

The usual procedure is to construct the door trim first and fit the baseboard and other woodwork to this. Of course, the door trim must be thick enough to accommodate whatever other woodwork meets it. In the traditional door trim there is often a separate block at the base of the trim. It is thicker than both the trim and the baseboard, allowing both elements to butt up against it. A similar block may also be located at the top of the door frame (to balance the lower block), where it may or may not serve to accommodate any woodwork. Where such blocks are used, the door trim is, of course, cut square to meet them. Where there are no blocks, the bottom ends of the side trim are cut square against the floor, but the upper ends are more likely to be mitered, unless the trim is made with extra molding applied over the main board. What happens above the door, particularly a front door, can be very involved, and extensive repair may require a qualified carpenter.

Removing Door Trim

Removing door trim is similar to removing window trim. Any applied molding must be removed first, then the trim, which is fixed to both the wall and the door frame (properly called the *jamb*). If the walls have been papered or painted up to the trim, run a razor blade down the edge of the trim against the wall so the paper or paint is not pulled off the wall when the trim comes away. When prying pieces loose, protect the wall carefully as you pry—a wood shingle works well.

Replacing Door Trim

When you replace door trim, attach the sides first. Note that the inside edge (nearest the door) is usually set back from the door frame 1/4 inch. Side and top pieces have to be cut to allow for that setback. Nail the trim to the door frame first, using 8-penny finishing nails. With a nail set, drive them just below the surface, so you can fill the holes with wood filler. Nail every 8 inches or so on both sides, and to help close the miter joint at the top corner, drive a nail in from the side, and another from the top (see Figure 57). Be sure to drill pilot holes in the corners first, or you'll run the risk of splitting the wood.

Most contemporary door trim available at lumberyards and home improvement centers is the "clam shell" pattern, and must be mitered (it doesn't meet neatly otherwise). For such cutting, a miter box, even the simplest kind, is indispensable.

If you are matching an older style of square-cut door trim, it is better, and easier, to cut the side pieces square at the top. The top piece will rest on them, reaching from the outside edge of one upright to the outside edge of the other. Even the square-end cuts are best done in a miter box with a fine-toothed saw. Molding applied over the outside edges must, of course, be mitered, but since the molding is much narrower than the actual trim, cutting a neat miter is quite easy. Whatever form of joint is used—square or mitered—if it is not

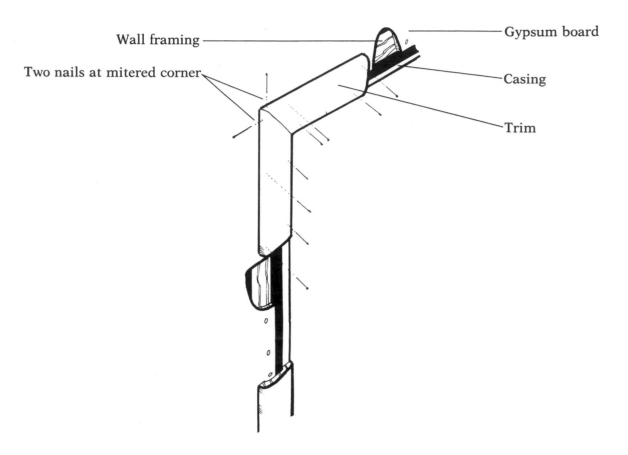

Wall framing

Gypsum board

Two nails at mitered corner

Casing

Trim

Figure 57 Nailing door trim

tight do not try to fill the gap with wood filler since it will inevitably dry and eventually fall out. Be patient and cut the joint again, even if you have to use another piece of wood.

BASEBOARD

Over the last hundred years, baseboard has gradually shrunk in size and design. Today, many new houses make do with a simple piece of one-by-four or even one-by-two with square edges, running around the bottoms of the walls. This is far different from the elaborate profiles found in older homes, where the baseboard (or *skirting board,* as it was frequently called) was often a foot high and made up of several different pieces of superimposed molding.

Even today's baseboards, however, usually manage to include a small additional piece at the bottom of the board, except perhaps in rooms fitted with wall-to-wall carpeting. Usually in a quarter-round shape, this piece is known as the shoe molding, and its function is to hide the gap between the bottom of the baseboard and the floor. Since the baseboard is installed on edge, it doesn't necessarily conform to any unevenness in the floorboards; the shoe molding is more easily bent to hide any such gaps.

Dealing with Shoe Molding   How the shoe molding is applied is important because improper nailing can cause it to split. When baseboards were higher, they were subject to shrinking across their height. That could result in the bottom of the baseboard lifting away from the floorboards. To hide such a gap, the shoe molding is nailed to the floor, not to the baseboard since it would only be pulled up along with the baseboard. Modern baseboards are much narrower and so shrink less; it has become common practice to nail the shoe molding to the baseboard. Nevertheless, it's preferable to nail it to the floor. Whatever you do, be consistent; do not nail the shoe molding to both the floor and the baseboard. If there is any shrinking, the shoe molding could easily split.

Properly applied shoe molding is mitered at inside corners, outside corners, and wherever lengths meet in the middle of a wall (see Figure 58). Butt joints at such midpoints tend to open up over time. The miter joint here is an *overlapping miter*—the miter of the second piece matches the angle of the waste cut off the first piece. Inside corners are sometimes coped—one piece is cut square and fitted right into the corner, while the end of the other piece is cut out to the shape of the shoe molding, so it fits tightly over the first. Coping is just as demanding as mitering—more so if the shoe molding has a complicated profile.

It's best to equip yourself with a simple miter box and learn to use it. The most important point is to hold the molding in the miter box tightly against the back of the box and keep the face side of the molding up—cut *into* the face side. Otherwise, a ragged edge may be the result where it cannot be hidden. Furthermore, since many corners are not perfect right angles, a miter box will enable you to adjust the cut so that it always bisects whatever angle the wall does make with its neighbor, whereas trying to cope to an odd angle is far more difficult.

Shoe molding need not be nailed more often than every 10 inches. Hold it tightly to the floor when nailing and nail it either to the baseboard or the floor, but not to both. Cutting inside pieces a little overlength (1/8 inch) and springing them into position will help ensure tight joints. Do not overcut outside corners. Lengths forming those corners must be cut exactly.

Removing Baseboard

When removing baseboard, use a sharp knife or a single-edged razor blade to cut the paint between any shoe molding and the baseboard. Remove the shoe molding first, prying it up from the floor or away from the baseboard, depending on how it was attached. Use a chisel or a pry bar and adequate protection—such as a wood shingle—next to the surface being pried against. That surface is often the wall, which is quite vulnerable above the level of the baseboard.

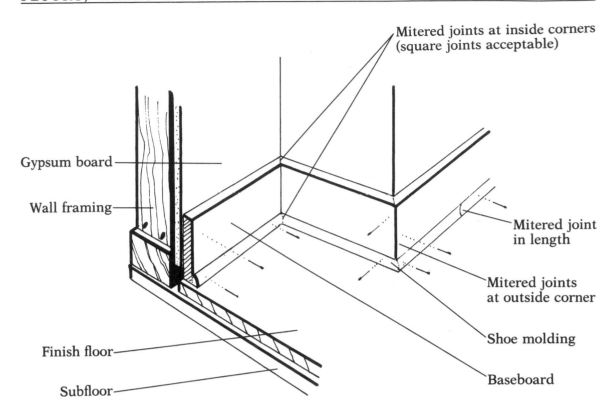

Figure 58   Baseboard nailing and jointing

If you try to remove a baseboard section completely, you're likely to break it off. Ease a little bit away at a time. Once you have pried a section of baseboard a little way away from the wall, insert thin wood wedges to hold the section just pried away, and work on the adjacent length. Apply more wedges and continue in this fashion until the whole length is freed. Undamaged sections thus removed, if carefully numbered and placed safely to one side, can be reused.

Any nails that remain in the removed sections should be pulled out with pliers through the back side of the board, not through the face side. Pulling from the face side will dent and blemish it. Make light pencil marks on the face side where nails were used so that you can be sure of renailing the baseboard to the studs and not just to gypsum board, which will not hold the nails.

Replacing Baseboard

Replacing baseboard is the same as replacing the shoe molding, described above. As just noted, the top edge of the baseboard must be nailed to the framing, not just the wall covering. Since the studs are usually placed at 16-inch intervals, starting from one of the corners, and since electrical outlet boxes are usually fixed to the sides of studs, you can usually locate the studs without too much difficulty. If you do have to guess, try nailing through the wall covering below the level of the baseboard. When you locate a stud, make a removable mark at that spot above the level of the baseboard. But don't make these trial holes too low or you'll find the wall plate, not the stud (see page 93).

One way to ensure tight outside corners is to end nail through both adjoining sections of baseboard in the same way the top corner of mitered door trim is nailed (see Figure 54). Here, too, drill pilot holes first.

When installing square-edged baseboard, inside corners can be finished with a simple butt joint. Cut both pieces square and butt one against the other in the corner. Molded sections may be coped. If you can trace the profile of the molding on the face edge of the adjoining piece, hold the coping saw at an angle very slightly greater than 90 degrees to the face of the molding and cut along the traced line. If your miter box is tall enough to hold the molding upright, it is easier to make a 45-degree cut. Then cut along the profile of the molding, holding the coping saw at an angle slightly greater than 90 degrees.

When two pieces of baseboard must be joined in the middle of a wall rather than in the corners, take care to make the joint over a stud so that the ends of both may be nailed securely. Miter them, if possible, rather than simply butt them together. Use the overlapping miter joint (Figure 58). If they should shrink a little, the mitered joint will not open up and produce a gap.

PART 3

CEILINGS

17

HOW CEILINGS
ARE CONSTRUCTED

TYPES OF CEILINGS

In a certain sense many ceilings are not actually constructed but are simply the result of other construction. What we think of as the ceiling is often no more than a surface covering over the bottom of the floor joists of the rooms above, or the underneath sides of the roof rafters. Sometimes, in fact, there is no covering at all, and the ceiling is "exposed"—you see the unadorned surface of floor or roof above. Understanding the "construction" of these ceilings only involves understanding how they were finished—with plaster, gypsum board, wood paneling, or various kinds of tile.

There is, however, one broad class of ceilings, known as *suspended ceilings,* that is quite different. The structure, surface covering, and finish of suspended ceilings are largely independent of any construction above, and thus they become a truly separate element in the overall makeup of a house. As such, chapter 18 is devoted to their description, repair, replacement, and construction.

The discussion that follows here, while primarily a description of different ceiling constructions, will prove invaluable for anyone considering altering, removing, or replacing an existing ceiling. Knowing what may be above your head will allow you to make sensible decisions when considering changes.

When the Roof Is the Ceiling

A roof is common to all houses, and thus all houses have at least one ceiling under the roof. The simplest is no ceiling at all—the underneath of the roof structure is the ceiling—called a *cathedral,* or *open, ceiling* (see Figure 59). The *rafters*—the long beams that support the roof covering—are left visible, together with any supporting members such as *collar beams* or *tie beams* (horizontal members joining opposite pairs of rafters at different points).

Older, heavier construction also involved such framing elements as king posts, queen posts, and purlins, but these have been largely replaced in contemporary construction by a *truss system,* using smaller pieces repeated at more frequent intervals. Truss construction does not generally lend itself to a cathedral ceiling, since the many trusses with their galvanized metal plates are not as pleasing to look at as the older beamed roofs.

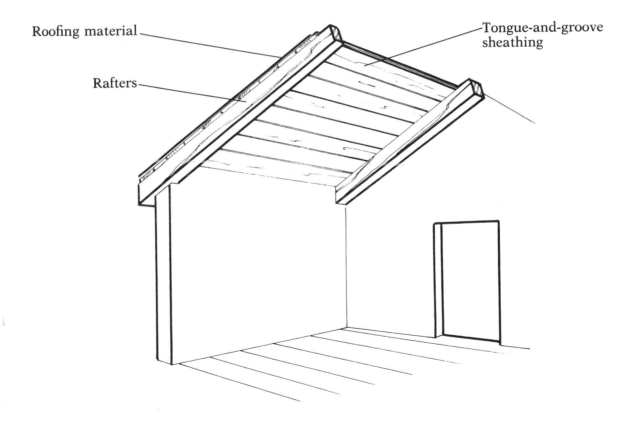

Roofing material

Rafters

Tongue-and-groove sheathing

Figure 59   Cathedral (open) ceiling

Exposed Board Ceilings   Trusses are generally used for roofs that span large distances with a relatively flat *pitch* (the measure of the roof's steepness). More steeply pitched roofs, and especially those made with larger framing elements, are likely to have fewer components. They are sometimes sheathed (covered on the outside) with tongue-and-grooved boards to which the roofing material itself is attached. These tongue-and-grooved boards, made with a V-joint at their edges, form a finished interior surface that is visible between the rafters and can be very handsome.

Many A-frame houses are constructed this way, the advantage being that every reliable inch of interior space is used, right up to the roof. Such a design, however, is not the most energy efficient. The roof must be insulated from the outside, which is sometimes difficult to accomplish, and the design usually results in a lot of overhead air space to which much of the house's heated air escapes.

Other Cathedral Ceilings   Cathedral ceilings are used even in buildings whose roofs are not sheathed with tongue-and-grooved boards. If rougher sheathing is used, the space between the rafters is insulated and finished from the inside (see Figure 60). Rafters must be deep enough to accommodate this extra material, although shallow rafters may be built up, provided the added weight can be safely supported by the roof structure as a whole. If the rafters are too small and too close together, however, such a procedure is really not practical. There is considerable labor involved in furring out the sides of every rafter, applying the necessary vapor barrier, insulation, and finished surface. Extreme care must also be exercised when painting or plastering the covering between the rafters.

But older buildings—structures such as converted barns or commercial buildings with handsome beams that are widely spaced—can make this kind of renovation a very attractive proposition.

The Condensation Problem   One of the problems with finished cathedral ceilings is how to deal with the inevitable condensation. When warm and often moist interior air hits a cold surface (such as a ceiling under the roof), the moisture in the air will condense. Most roofs are built to provide adequate ventilation between their subsurface and the ceilings below them.

One way to achieve a cathedral ceiling and yet provide such ventilation is either to construct a small horizontal section of ceiling just under the *ridge* (the peak of the roof), or to use existing collar beams to build a ceiling that will leave a space above them but below the top of the roof. Such a space will allow the air to circulate and thus prevent the accumulation of moisture that can condense.

The space alone is not enough, however. Louvered vents must be placed at each end of this miniattic. A prefabricated roof ridge-vent is usually installed, and a fan to encourage adequate air circulation. Insulation, which is normally located in the space between rafters, must be removed from the rafters in the miniattic and replaced between the collar beams that now form the ceiling.

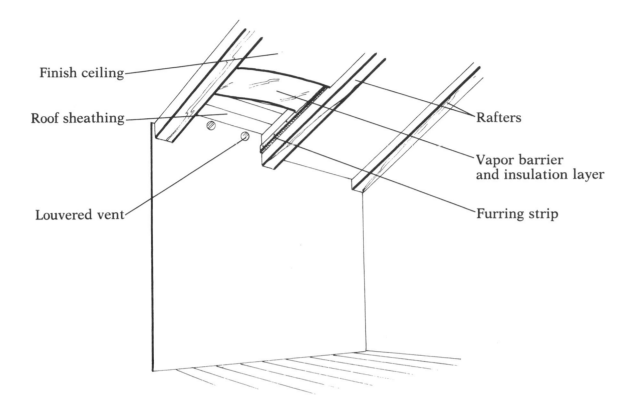

Finish ceiling

Roof sheathing

Rafters

Vapor barrier
and insulation layer

Louvered vent

Furring strip

Figure 60 Finished cathedral ceiling

Flat Ceilings

The next type of ceiling is a flat ceiling above which there are no rooms, just a small unfinished space beneath the roof (see Figure 61). Since the framing for such a ceiling has only to support its own weight plus the weight of the ceiling's finished surface, it can be much lighter than any other horizontal framing in the house.

Even if the space above the ceiling is accessible, it is inadvisable to walk on the ceiling. There is always the risk of falling through the ceiling material or breaking the supporting members. These beams may only be strong enough to support the ceiling; they may also act as tie beams, keeping the outer walls from being pushed outward by the weight of the roof.

Ceilings Under Floors

The third type of ceiling is one applied to the undersides of structures built to support floors above. The only difference, if any, between these structures and regular floor framing

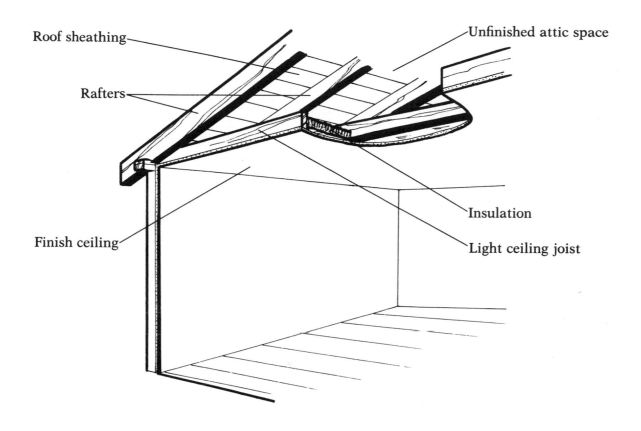

Roof sheathing

Rafters

Finish ceiling

Unfinished attic space

Insulation

Light ceiling joist

Figure 61   Flat ceiling under roof

structures is that the bottom surface is built carefully so that it is level enough for the applied ceiling to be perfectly flat. In the upper stories, bearing walls act as girders, which are in turn supported beneath the floor on which they rest either by further bearing walls or by girders or piers in the basement (see page 9).

Since such ceilings are floored over from above, any wiring or ducting of services (plumbing, heating, air-conditioning) that requires rerouting or repairing may have to be reached from the ceiling side rather than the floor side. Before performing any major surgery on such a ceiling, you should be familiar with whatever utilities are housed in the ceiling and make provision for future access, should that become necessary.

CEILING SURFACES

Any finish material that can be used for an exposed ceiling can be used for a flat ceiling, whether or not that ceiling has a room above it. In the Middle Ages this covering, of wood or plaster, was applied by tradesmen called ceilers, whose job of "ceiling" also extended to walls. The word *ceiling* is no longer a verb and is now limited to the covering above our heads.

Wood-paneled ceilings are now rare, presumably because of expense—there is no structural reason not to use wood. At the beginning of the present century most self-respecting houses were built with plastered ceilings applied over a framework of lathing (as described in pages 94–95). Plastering a ceiling was one of the most difficult facets of house construction.

The introduction of plaster in sheet form, as gypsum board (widely in use since World War II), made lath-and-plaster ceilings an unnecessary expenditure of time and labor, especially when all that was needed was a plain, flat surface. Indeed, gypsum board, if properly applied, will last longer without cracking (and possibly falling down) than nearly any solid-plaster ceiling. Gypsum board does not, however, allow for the now rarely seen molded-plaster roses and decorated friezes, the repair of which is difficult, expensive, and best left to professionals.

The use of patterned stamped metal, usually tin, as a ceiling covering was also popular at one time, and originated as a quick and cheap substitute for patterned plaster ceilings. It was especially common in institutional buildings. While the material is available today, it is designed primarily for restoration work. Tin ceilings are generally inappropriate in the modern homes of today.

Applying the Finish

Since rafters and floor joists, as well as ceiling joists (those supporting no roof or floor above), are all generally spaced 16 inches apart, just like wall studs, the procedures for locating them and applying gypsum board to them are the same as for walls. Gypsum-board ceilings may be taped and spackled (as described on pages 111–13); given additional texture by the application of layers of joint compound, troweled or sponged; painted (as described on pages 117–25); or covered with ceiling paper or even patterned wallpaper (as described on pages 130–38).

Installing a gypsum-board ceiling requires no more skill than installing a gypsum-board wall, but it is far more laborious. A helper is essential to put up the board. Work from a secure scaffolding while you install and tape. You can rent scaffolding; consider renting a gypsum-board lift that holds the sheet flush to the rafters while you and your helper nail it in place.

Covering a ceiling with wood—in the form of solid boards or sheet paneling—follows the same principles as wall paneling (described in detail on pages 146–50).

Ceilings are rarely finished with ceramic tile, but use of special ceiling tiles made of a variety of other materials is very common. Tiling a ceiling is discussed at length in the following chapter on suspended ceilings. Suspended ceilings are invariably made with various forms of tile, and the techniques involved apply generally to the tiling of any ceiling.

In the area of novelty ceiling treatments, the most common is the use of imitation wood beams. Usually made of fiberglass, they are designed to impart the charm of older post-and-beam structures. Light in weight, they are applied easily with adhesive and a few nails. While such beams can be reasonably realistic, they must be applied with some understanding of the roles that the original beams played in the structure of the house. Otherwise, the effect can be ludicrous.

Working on a Ceiling   It is almost always easier—and preferable—to do the work on a ceiling, whether it be a simple repainting or a complete renovation, before doing any work on the walls or the floor. This is partly so you don't have to worry about paint or other spills. More important, however, is that work on ceilings requires elaborate support structures, such as scaffolding, which can be more easily erected if walls and floors are unfinished. The scaffolds can be tied into wall framing and ladders nailed temporarily to floor surfaces, if necessary.

18

SUSPENDED CEILINGS

A suspended ceiling can be defined as any ceiling whose finished surface is neither part of, nor directly applied to, the structure of the roof or of the floor above. A suspended ceiling, therefore, can be removed without any effect on the building's structural integrity.

TYPES OF SYSTEMS

Within this broad definition there are various suspension systems that differ quite radically from one another. The first difference has to do with how far the ceiling is suspended from the structure above. It can be suspended as far as several feet or as little as the thickness of a one-by-two furring strip.

The second difference has to do with what kind of suspension is used. The suspension framework can be of metal or wood hung from, or otherwise attached to, the unfinished ceiling above. This framework supports panels or tiles of different sizes and various materials.

Finally, there are differences in the actual material out of which the tiles or panels are made. These may be designed for practical purposes, such as sound absorption or light reflectance, or purely for appearance. The material ranges from lightweight fiberglass to plastic or other compositions.

Why a Suspended Ceiling?

A suspended ceiling can solve a variety of problems. Such a system is often the most convenient way of dealing with obtrusive utilities, such as pipes, wiring, and ductwork, that cannot be otherwise hidden. A suspended ceiling is often the quickest way to create an acceptable finish overhead when there is no ceiling or when the existing one is in bad condition. Suspended ceilings can often provide for special needs not easily accommodated by conventional ceilings, such as hidden or recessed lighting or special acoustic effects.

The Packaged Ceiling

You can fabricate your own suspension system, but most such ceilings are bought in kit form. They're designed so they can be installed in almost any location. Furthermore, the tiles and panels designed to be used with such systems are often replaceable and sometimes interchangeable. That offers you endless design possibilities; you can develop many unique applications for individual situations. The principles outlined in the next section should serve to make installation of almost any system comprehensible and within reach of the average do-it-yourselfer.

INSTALLATION OF SUSPENSION SYSTEMS

One of the reasons for installing a suspended ceiling, as noted, is to hide pipes and ducts that run across the ceiling rather than through it or within the walls. This is particularly common in a basement that was not originally designed to be living space but which is now being turned into a finished room. Such obstructions will usually determine the height at which the ceiling can be hung. If possible, the ceiling should be hung low enough to cover all pipes and ducts. However, many building codes stipulate minimum height requirements. Sometimes these require an overall minimum clearance, though they may permit partial lower areas if they occupy less than a certain percentage of the ceiling's total area.

A suspended ceiling with panels supported in a grid system will be much easier to install if the entire ceiling can be hung at the same height. If you have to construct the ceiling at different levels, or even accommodate only one low section (perhaps to box in an especially large duct), it will be easier to use a furring-strip system (see below). These tiles are installed over their supporting framework rather than being supported in a grid system, and it's far easier to build that framework (usually furring strips) around obstacles.

Acoustical Tile Ceilings   *Acoustical tiles,* sometimes called *interlocking tiles,* provide increased levels of sound absorption, thermal insulation, and fire resistance. They are usually washable, relatively inexpensive, and generally easy to install, requiring only one special tool, a staple gun.

185

You may occasionally want to tile over an existing ceiling in good condition—a level, unbroken surface such as gypsum board—for decorative or acoustical reasons. In that case, you can cement the tiles directly to the ceiling. More commonly, you'll staple the tiles to a framework of furring strips made from lengths of one-by-three (see Figure 62). (Acoustical tiles can also be installed in a suspended grid system.) The edges of the tiles lock into neighboring tiles and form a secure, level surface.

Laying Out the Job   The arrangement of the furring strips is the first job to be done and requires careful planning. Most tiles are 12 inches square (although tiles measuring 12 by 24 inches and 16 by 16 inches are also available), and the furring strips to which they are stapled must be spaced accordingly. Unless the ceiling measures an exact multiple of 12 inches in both directions, the border tiles will have to be cut, and the supporting furring strips placed accordingly.

The best way to position the furring strips is to make an exact plan of the ceiling on graph paper and pencil on it the location of the strips. Whether the furring strips are to be nailed directly to the ceiling joists or nailed through some existing surface such as plaster or gypsum board, the strips should run at right angles to the ceiling joists. Design the layout so that the border tiles measure about the same width on opposite sides. If the border tiles turn out to be less than half the width of a tile, move the whole grid over half a tile. The border tiles should now be more than a half tile width.

Finding the Joists   If the joists are not visible or accessible from above, and if there is no other way of determining their location, you can find them by maneuvering a piece of wire, such as a straightened coat hanger, through a small hole drilled in the ceiling. Bend it from one side to the other. Since most joists are positioned at intervals of every 16 inches, you

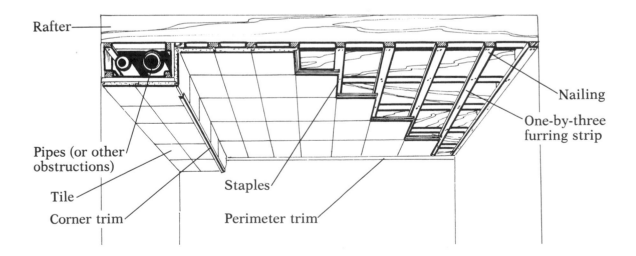

Figure 62   Acoustical tile ceiling installed on furring strips

should never be further than 8 inches away from one. Having located one joist, you should find the remaining ones by measuring out in increments of 16 inches.

Putting Up the Furring   Following your plan, nail the furring strips with two 8-penny nails at each juncture with a joist. Use a level to ensure that the ceiling is flat. If it's not, use shims behind the furring strips to level them. Install furring strips around any obstructions as well. You can always increase the depth of the furring strips by laying a second set at right angles over the first. This can accommodate pipes, wiring, or conduit that may be slightly deeper than the existing surface. Larger obstructions should be completely boxed in, providing furring strips over the boxing, to which to attach tile. Although it involves considerably more work, you can suspend the furring strips by constructing a support system of almost any depth, to which the furring strips can be nailed. Such a system will create space above the tiles to accommodate recessed light fixtures.

When the furring strips have all been installed, snap a chalk line down the middle of the first strip butted against one wall, to provide a guideline to which the tiles can be aligned. At both ends of an adjacent wall measure out the width of the border tile to be installed along that wall and snap a second chalk line *across* the furring strips. You'll start tiling from the corner where the chalk lines meet. Make sure the lines are perpendicular to each other using the 3-4-5 right triangle method described on page 37. Resnap the second line, if necessary.

Putting Up the Tile   Open the boxes containing the tiles 24 hours before you start to allow them to become acclimatized to the room's temperature and humidity. To avoid soiling the tiles, it helps to rub talcum powder on your hands. The tile boxes usually carry instructions—be sure to read them before you start work. Cut the border tiles running along both marked walls first, using a utility knife and a straightedge. Staple them to the furring strips so that the flanged sides, through which the staples are inserted, face into the center of the room. Allow a 1/8-inch gap between the tile and the wall for possible expansion, but make certain the flanged edge is exactly flush with your squared chalk lines. A small error here will compound itself later. Work into the room from these two sides, taking care to fit the next tile over the flanges of the previously laid tiles. Most such tiles require three staples along one flange and only one on the other, so the neighboring tile can be fitted over this edge.

When the last two walls are reached, and the border tiles for these have been cut and installed, a trim strip is nailed up around the perimeter. It supports the tiles, covers the expansion gap, and hides any irregularities. If the trim strip is to be painted or stained, do this before nailing it in place.

Suspended-Grid Systems

The panels in a suspended-grid system can easily be removed, allowing convenient access to any pipes, wiring, or ducts that may be located above the panels. But even if there are none of these, the grid system must still be installed at least 3 inches below the surface of the existing ceiling (or below the bottom of the ceiling joists if the ceiling is unfinished); otherwise, it would not be possible to insert the panels up through the grid to drop them in place. If you plan and plot your layout carefully, installing a suspended-grid tile ceiling is quite easy.

Laying Out the Grid   Just as with acoustical tiles, planning the layout of the grid carefully is the most important first step. *Runners*, supported at each end by wall molding, are arranged at right angles to the ceiling joists, from which they are hung by *suspension wires* (see Figure 63). Between these runners, *cross-tees* are installed. Together with the runners, these cross-tees provide the support necessary for the panels.

Start by making a plan of the ceiling so you can decide exactly where the runners will be located. Panels are made in various sizes (although 24 by 48 inches is by far the most usual size) and should be installed so that their short sides rest on the runners, and their long sides rest on the cross-tees. Since many rooms have the ceiling joists running across the shortest dimension, and since the runners should be installed at right angles to the joists, the panels will also run across the room, usually the best arrangement. Place the runners and the cross-tees so that the panels at each side and at each end of the room are equal in length and width, respectively.

Condensation from sweating pipes, and even humidity, can cause metal suspension systems to rust. This can, in turn, stain ceiling panels, so any pipes that are hidden behind the ceiling should be wrapped to prevent condensation.

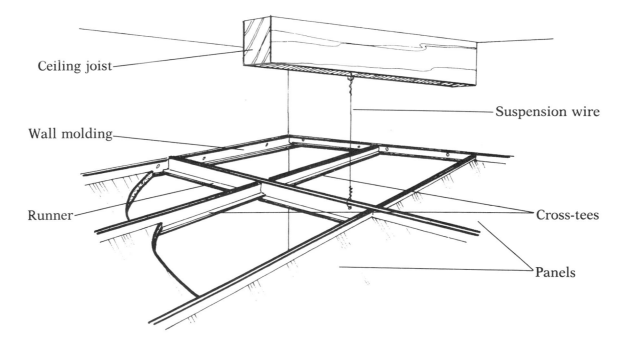

Figure 63   Suspended grid system

Putting Up the Grid   When you've made the layout, snap chalk lines around the walls at the height at which the ceiling is to be installed. Even if the ceiling is not being lowered an amount determined by obstructions, remember to leave a minimum of 3 inches above the grid system so that the panels can be inserted through the metal framework. Similarly, if you plan to install recessed light fixtures, leave enough space for them.

On these chalk lines, nail up the wall moldings. Now fix a stretched string from one end of the room to the other at the level of the wall molding and at the exact location of the first runner at one side. Stretch a second string at right angles to the first at the location of the first cross-tee. Now you can install the first runner, cut to length so that the joint for the first cross-tee is directly at the point where the two stretched strings intersect. This will ensure that the first border panel is the proper width. The remaining runners are installed by measuring from the first, and when all are in place, the cross-tees are installed on the runners.

The whole system of runners and cross-tees is supported at the sides of the room by the wall moldings and, across its open space, by hanger wires attached to the joists. Fix screw eyes to the bottom of the ceiling joists immediately over the runners, then bend the hanger wires at just the right height so the entire system of runners is perfectly level. Test with a level to be sure—you can adjust the hanger wires if necessary. The cross-tees are simply laid in place.

Inserting the Panels   The panels used in such grid systems are made of various materials, but all are very easy to cut to size, where necessary. It is very easy to damage panel edges, so work carefully, especially when inserting them through the grid system.

Special panels called *light lenses* can be used in place of regular opaque panels, allowing light fixtures to be installed entirely within the suspended ceiling. As an alternative, regular recessed fixtures can be installed in the panels themselves simply by reinforcing the opening cut for the fixture. Do this with furring strips that run the length of the panel and rest on the grid.

MAINTENANCE AND PRECAUTIONS

Some tiles and panels are made with asbestos, fiberglass, and other potentially harmful substances. Be sure you are aware of the composition of any material you work with and wear masks and other protective clothing where appropriate. Avoid installing asbestos tiles.

Excessive humidity can cause some materials to warp. Fiberglass, however, does not warp, and if you cannot provide adequate ventilation, try installing fiberglass panels.

Under normal conditions and with regular cleaning and occasional repainting, both tiles and panels will last for years. Note that painting can affect the acoustical quality of tiles by filling the holes or fissures—use a well-thinned paint.

19

CEILING PROBLEMS

STRUCTURAL PROBLEMS

There are as many potential structural problems with ceilings as there are different kinds of ceiling structures. Identifying some of the more serious ones, however, may prevent vain attempts to repair what are wrongly perceived as surface problems (see Figure 64).

Sagging Ceilings

A sagging ceiling can indicate a serious structural problem. A solid plaster ceiling, and sometimes a badly fixed gypsum-board ceiling, may belly out for reasons that are *not* structural, but usually these faults are the result of structural defects. If the roof leaks, the damage will inevitably show up first in a ceiling. Water can even collect behind plaster—without passing through—and eventually force the plaster to collapse by its water-logged weight.

Insufficient Support

Sagging ceilings may also indicate insufficient support. Older houses may have had walls removed that should not have been—take a close look at the floor plan to see if any anomalies are apparent. Apart from sagging ceilings, clues that walls have been removed are empty notches in exposed beams, patched floorboards, uneven wall spaces between windows, and

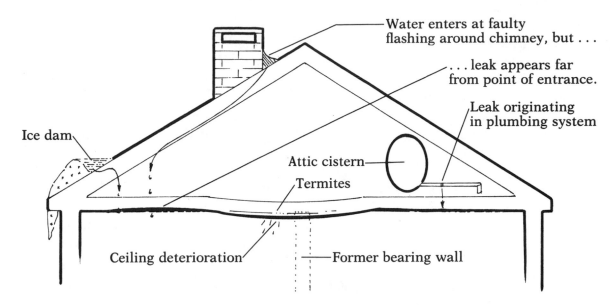

Figure 64 Clues to ceiling damage

interrupted or unfinished sections of molding around floors and ceilings.

Even if no original support has been removed, older houses may develop sagging ceilings because the original support was insufficient. Massive beams that may have seemed sufficient to the builders at the time sometimes sag with the passage of years. This often happens to ceilings below rooms that have been converted from previously empty attic space, especially if these rooms now hold unplanned-for bookcases, pianos, or other heavy furniture.

Bowing of the Walls

A somewhat related effect with a different cause is produced when the walls of old houses (or even poorly built new houses) start to bow outward under pressure from the roof. This usually indicates that the collar beams or tie beams (whose job is to hold the sides of the house together) have failed or are insufficient. This may also be the result of an ill-considered renovation.

Obtain the advice of a qualified building engineer or contractor if your ceilings sag and walls appear to have been removed or relocated or if exterior walls are not plumb and the roof ridge also sags. The chances are that the deformation has stabilized and the house is structurally secure, and you may then fix the ceilings without fear of wasted effort. But if remedial construction is necessary, it is better to attend to this before patching or replacing ceilings.

Other Structural Problems

There are structural problems not part of any overall systems failure that can affect ceilings. Since ceilings are often formed on the underside of structures that have other, primary functions—serving as the floor of the room above, for example—work that affects these areas will inevitably have an impact on the ceilings. Floor repairs involving the laying of additional levels of wood or tile, rewiring projects that involve drilling through floor joists, and even stairway repairs or alterations can cause direct damage to ceilings even though they may not involve actual ceiling work.

When attic floors are insulated, there is the risk of damaging the ceilings below. Repairing these can produce a terrible mess if loose insulation was used without any underlayment between the ceiling joists.

Insect Damage

Last, you should suspect insect damage—from termites, carpenter ants, or powder-post beetles—if applied ceiling material such as gypsum board or wood paneling is hard to attach securely. If ceiling joists are weakened too much, they may collapse, but you should have plenty of warning. Long before anything like that happens nails will lose their holding power, and it will be difficult to drive in new ones securely.

Insect damage, which shows itself quickly in uncovered areas because of telltale piles of dust or exit holes in joists and sills, is hard to detect if covered by a finished ceiling. Regular pest inspections followed by extermination programs when called for is the best way to prevent serious damage of this nature.

LEAKS

Leaks in roofs and plumbing systems can cause serious damage to ceilings and must be attended to before the ceiling can be repaired. This is easy to say but not always so easy to do. Finding the cause of the leak can often be more difficult than repairing it. An obvious first step is to determine whether the leak originates from the outside or the inside of the building.

Outside Leaks

If it's been raining for a while, it is logical to suspect the roof. But leaks from the outside can appear long after precipitation, particularly in winter. Melting snow or ice that may have formed weeks earlier might be the cause. Ice dams that form at the edges of roofs, especially on overhangs that receive no heat from the house, can trap water that has melted and run down from higher areas. This accumulated water can back up under the shingles until it finds its way through the roof. If you observe any such accumulations of ice around the edges of the roof, they should be chipped away, taking care not to damage the roofing material. Allow the trapped water to drain away.

Should gutters and downspouts become blocked—either with debris or ice—they may also cause ice dams. Maintain clean gutters, ideally inspecting them every fall before winter sets in, to ensure that they're clean and in good condition. Other areas that can become blocked and cause ice dams are valleys between adjacent roof slopes and around chimneys.

Roof Leaks

Water damage caused by roof leaks themselves may occur a long way from the actual leak, since water entering through roofing may trickle down the underside of rafters and other roof members before hitting the ceiling. If you can easily locate the leak from the exterior of the roof, you may be able to repair it on the spot. More likely, however, you'll have to inspect the roof from beneath, using a flashlight to trace the course of the water from the point at which the ceiling is damaged back to the point of water entry. Once you've located the weak area in the roof, poke a thin nail through from the inside so a helper on the roof can spot it. Now you can properly repair it from the exterior.

Use a caulking gun around exterior window and door trim to prevent rain and moisture from entering the house and finding its way to various ceilings. That will also improve the house's insulation.

Inside Leaks

If it hasn't rained for some time, it is logical to suspect that the leak is inside. The easiest way to find out is to cut off the water supply to all points above the appearance of the water damage and drain the system to this point. This is most easily done by closing the appropriate shut-off valves and leaving the drains open. However, there may be other plumbing parts that are not so easily drained, such as sprinkler systems, which must be inspected directly. Another potential site may be pipes that are sound in themselves but subject to condensation—these can cause moisture that can damage ceilings.

If the leak is in the plumbing, you may need a professional plumber to fix it. A professional may also be able to locate a plumbing leak if you can't.

Repairing Water Damage

The effects of water damage can be minor, requiring little more than a paint touch-up. But they can also be severe, calling for complete replacement of the surface material (detailed in the relevant preceding chapters). One special caution when you're dealing with a ceiling that has suffered water damage: Turn off the electricity if there is wiring or an electrical fixture in the ceiling.

Appendix:
Selected Product Ratings

The following Ratings give Consumers Union's evaluations of some products used to finish floors, walls, and ceilings. The Ratings of the individual brands are based on CU's laboratory tests, controlled-use tests, and/or expert judgments. To minimize loss of usefulness because of periodic reformulation of products, we've included Ratings of only those products that we've tested relatively recently. Therefore, this appendix is by no means a comprehensive guide to every building supply you'll need. But it can give you a good idea of which clear wood finishes, spackling compounds, interior latex paints, and high-gloss enamels are likely to perform best.

Although the Ratings are not an infallible guide, they do offer comparative buying information that can greatly increase the likelihood that you will receive value for your money. You may be tempted to purchase whatever brand appears at the top of the Ratings order. Resist the impulse if you can. Before you read the Ratings, read the text and Guide to the Ratings section that precedes each chart to determine which product is best suited to your needs, and then read the notes and footnotes. In those sections you will find the features, qualities, or deficiencies shared by the products in the test group and the basis on which the order of the Ratings was decided.

For many purchases, particularly when you're buying large quantities, it's worthwhile to comparison shop. Call several stores to find out what brands they carry at what price.

CLEAR WOOD FINISHES

For refinishing a floor, look for a varnish that dries quickly and that scored well in our tests for hardness, flexibility, and resistance to wear and water. A top-rated high-gloss varnish would be fine if you want a very shiny finish. The satin or semigloss varnishes we tested came close to providing the natural wood look of penetrating oils, and were generally tougher than their high-gloss counterparts. Penetrating oil protects wood by soaking into the pores. It does not leave a surface coating, as varnish does, and as a result does not do much to protect the wood from abrasion.

Guide to the Ratings

Listed by types; within types, listed in order of estimated quality.

Ease of application. With the varnishes and lacquers, the lower the score, the more the brush dragged when we applied a second coat. The penetrating oils were all relatively easy to apply with a brush or a rag, but the excess oil must be wiped off. We debited the oils for that simple, but messy, extra step.

Resistance to sagging. Because these products are thin, they sagged and ran considerably, even on horizontal surfaces. Second coats were especially likely to sag, since the wood absorbed less finish the second time around.

Drying time. The more quickly a finished surface dries **tack-free**, the less unsightly dust and lint it picks up. Some finishes, including the lacquers, were tack-free in about half an hour; some needed almost a day. To dry **hard**, surfaces needed anywhere from about an hour to nearly two days. Surfaces should be left to dry hard between coats.

Ease of sanding. These products should be sanded between coats. Most varnishes tended to gum up and clog fine sandpaper. Even the finest sandpaper leaves scratches when it's unevenly clogged. As you might expect, oils were generally easier to sand. That's because you are sanding the wood itself along with the finish.

Initial color. Some of the finishes looked surprisingly dark, almost amber, rather than light. On dark or stained wood, that may not matter. But to keep the whitish look of birch or pine, shun finishes with a low score here, especially if the job calls for several coats.

Hardness. The judgments here are important if you're refinishing a floor. The higher the score, the better the finish resisted scratches and abrasion.

Flexibility. Even the hardest finish must bend and stretch with the surface it covers; a brittle finish tends to crack or chip. We coated tin-plated panels with each finish, let them age, then bent them and checked for cracks and peeling.

Resistance to wear. We applied each finish to oak flooring that we laid in heavily used corridors at CU. Most varnishes and lacquers held up quite well. We did not score the penetrating oils for resistance to wear. We did note, however, that oiled and uncoated panels quickly became soiled despite frequent moppings.

Resistance to blocking. A clear finish may remain somewhat tacky long after it appears

Ratings

As published in an August 1987 report.

Clear wood finishes

Legend: Better ← ● ◐ ○ → Worse

Rating categories (columns, left to right):
- Ease of application
- Resistance to sagging
- Drying time, tack-free/hard
- Ease of sanding
- Initial color
- Hardness
- Flexibility
- Resistance to wear
- Resistance to blocking
- Resistance to water

Brand and model

HIGH-GLOSS VARNISHES AND LACQUERS

- Flecto Varathane Professional 900
- Benjamin Moore Benwood One Hour 42000
- Sears Open Hearth Polyurethane Gloss Varnish 60224
- Flecto Varathane Liquid Plastic Clear Gloss 90
- Pittsburgh Rez-Polyurethane Gloss Varnish 77-5
- Minwax Polyurethane Clear Gloss Finish
- Deft Clear Wood Finish Gloss [1]
- Deft Wood Armor Gloss [2]
- Tru-Test Woodsman Polyurethane Varnish Gloss 012
- UGL Gloss Zar Polyurethane
- McCloskey Dura-Fame Polyurethane Varnish High Gloss 1090
- McCloskey Gym Seal
- Red Devil Polyurethane High Gloss 70
- Benjamin Moore Benwood Polyurethane High Gloss 42800
- Deft Defthane Polyurethane Finish Clear Gloss 1
- Valspar Gloss Finish Clear 10
- Pittsburgh Rez-Varnish Gloss Oil 77-4
- Valspar Polyurethane Liquid Plastic High-Gloss 20
- McCloskey Heirloom Clear Varnish High Gloss 0092
- Glidden Woodmaster Polyurethane 81 High Gloss
- Red Devil Clear Varnish High Gloss 50

SATIN AND SEMIGLOSS VARNISHES AND LACQUERS

- Flecto Varathane Professional 1100
- Red Devil Polyurethane Satin 71
- Minwax Polyurethane Clear Satin Finish
- UGL Satin Zar Polyurethane

Table of wood finish products with ratings (filled, half-filled, and empty circles indicating performance levels; column headers not visible on this page):

Pittsburgh Rez-Polyurethane Satin Varnish 77-9
Deft Wood Armor Satin [2]
Tru-Test Woodsman Polyurethane Varnish Satin 039 [3]
Valspar Polyurethane Liquid Plastic Satin 21
Deft Defthane Polyurethane Finish Clear Satin 2
Flecto Varathane Liquid Plastic Clear Satin 91
Pittsburgh Rez-Varnish Satin Oil Clear 77-7
Benjamin Moore Benwood One Hour Clear Finish Low Lustre 42100
Glidden Woodmaster Polyurethane 82 Satin Sheen
Deft Clear Wood Finish Semigloss [1]
McCloskey Dura-Fame Polyurethane Varnish Satin Sheen 1075
Sears Open Hearth Polyurethane Satin Varnish 60214
Benjamin Moore Benwood Polyurethane Low Lustre 43500
Red Devil Clear Varnish Satin 51
McCloskey Heirloom Clear Varnish Semigloss 0060
Valspar Chippendale Varnish Satin Finish 11

PENETRATING OILS

Formby's Tung Oil Finish High Gloss (16 oz.)
Formby's Tung Oil Finish Low Gloss (16 oz.)
Minwax Antique Oil Finish
Minwax Tung Oil Finish
Gillespie Tung Oil Finish Satin Gloss
Hope's Tung Oil Varnish
Minwax Wood Finish Natural 209
UGL Wipe-On Zar Clear Tung Oil Finish Original Semigloss [3]
McCloskey Tungseal Tung Oil Varnish Clear 931
UGL Wipe-On Zar Clear Tung Oil Finish Satin [3]
Gillespie Tung Oil Finish Low Gloss
Flecto Varathane Natural Oil Finish Clear 66
Deft Deftco Danish Oil Finish Natural
Watco Wood Floor Finish Natural
Watco Danish Oil Finish Natural

• The following product is listed separately because of wrinkling in laboratory tests, preventing measurement of some properties. The product offered no problem on wood.

Hope's 100% Tung Oil

[1] Lacquer [2] Water-based (latex) [3] According to the label, not recommended for floors

to be dry—especially in hot weather. To test for this drawback, known as "blocking," we applied each finish to a pair of standard paint-test charts, and we aged the finishes for at least a month. Then we placed each pair of charts face to face and applied about six pounds per square inch of pressure overnight.

Resistance to water. We spotted test panels with water, covered the spots, and checked for water damage. Most recovered fully overnight. Those that didn't were permanently clouded.

SPACKLING COMPOUNDS

In the past few years, a new sort of spackling compound has come along that's superior to its predecessors in many ways. We refer to the type as "lightweight," since a container of one of these products feels almost empty when you heft it. The new compounds are filled with air in tiny glass spheres. If you've had difficulty getting good, smooth patches, try one of the lightweight compounds. We think you'll find it easier to get good results.

If you're filling a very large hole or areas subject to wear and tear, you may want a tougher material than a lightweight. You're best off with one of the powdered compounds. Powders did well in deep fills, and most were so hard that you'd have to be careful not to bend a nail as you hammered it. (Try drilling a pilot hole before driving the nail.)

The powdered products have a small extra advantage: They store well if kept in a dry place. The puttylike premixed compounds may tend to dry out and harden once their containers have been opened.

Wear a dust mask that covers your mouth and nose when mixing powder or sanding a patch. And, of course, keep spackling compound out of children's reach.

Guide to the Ratings

Listed in order of estimated quality, based equally on convenience; resistance to sagging, cracking, and shrinking; toughness; and ease of sanding. Products judged about equal in quality are bracketed and listed alphabetically.

Type. L = lightweight paste. **P** = powder (must be mixed with water). **RP** = regular paste.

Convenience. In judging how easy each product was to apply, we looked at ease of transferring it from the container to the work surface, spreading and smoothing it, and filling gaps. We also looked at any tendency to stick to the spreading knife or pull away from the edges of a gap. Because the powders have to be premixed, their scores were all lowered one notch.

Resistance to sagging. Products that sag under their own weight will require users to fill some holes in multiple layers. The best performers here sagged not at all, so they would fill even fairly large holes in one application. **Resistance to cracking, shrinking.** Some products tend to crack and shrink when their water evaporates. Those, too, will require you to apply multiple layers.

Ratings

Rating	Symbol
Excellent	◉
Very good	◓
Good	○
Fair	◒
Poor	●

Spackling compounds

As published in a **July 1987** report.

Brand and variety	Type	Convenience	Resistance to — Sagging	Cracking	Shrinking	Toughness	Ease of sanding
Elmer's Redi-Spack Lite	L	◉	◉	◉	◉	○	○
Muralo Spackle Lite	L	◉	◉	◓	◒	○	◉
Red Devil Onetime Spackling	L	◉	◉	◉	◉	○	○
Servistar Spackling Compound	P	○	○	◉	◉	◉	◓
Bondex Spackling Powder [1]	P	◓	○	◉	◉	○	◓
Bondex Super Patch [2]	L	○	◉	◉	◉	○	○
DAP Fast'n Final	L	◓	◉	◉	◉	◒	○
Savogran Level-Best	P	○	◓	◉	◉	◉	◒
Sears 59325	P	○	○	◉	◉	◉	○
UGL 222 Lite	L	◓	◉	◉	◉	○	◒
Durabond Spackling Powder	P	○	○	◉	◉	◉	◒
UGL 222 Spackling Paste	RP	◉	◓	◒	●	◓	◒
Bondex Spackling Paste	RP	◓	●	◒	◒	◓	◒
Muralo Spackle	RP	◓	○	●	◒	○	◓
Sears 59334	RP	◉	○	●	●	◓	○
Durabond Spackling Putty	RP	◉	●	●	●	◓	◓
Elmer's Redi-Spack	RP	○	○	●	◒	◓	○
Red Devil Spackling Compound [3]	RP	◓	◒	○	●	◓	◒
Savogran Presto Patch	RP	◓	●	◒	◒	◓	○
M-H Ready Patch	RP	◓	●	◒	●	◓	◒
DAP Vinyl Spackling [1]	RP	◓	●	●	◒	◓	○

[1] Brittle when dry; split when a nail was driven into patch.

[2] Somewhat dry; tended to roll off spackling knife. Needed addition of a little water to improve workability.

[3] Labeled "lightweight formula," but judged not a true lightweight in weight and performance.

Toughness. Important for holes as deep as a half inch or for an area that must support a heavy picture.

Ease of sanding. Important if you're patching large areas, which can be hard to trowel off smoothly.

INTERIOR LATEX PAINTS

Painting may not be the romantic pursuit the industry wants us to believe, but it's a necessary part of owning a set of walls.

The Ratings will steer you to the brand and color that can best help you carry out your decorating plans. Although we can't recommend one brand for top performance in all areas and in all colors, we can single out specific brands for specific colors. The lists that follow will start you in the right direction. The paints we've listed combine good hiding power with other useful attributes. They resist dirt and hold up well to scrubbing, they go on without much spattering, they resist fading, and they stay tack free. Call several different suppliers to see which brands they carry at what price.

White
Pratt & Lambert low-luster finish
Kelly-Moore flat finish
Benjamin Moore flat finish

Blue
Pratt & Lambert low-luster finish
Pittsburgh Manor Hall low-luster finish
Kelly-Moore flat finish

Pink
Pittsburgh Manor Hall low-luster finish
Pratt & Lambert low-luster finish
Tru-Test EZ low-luster finish
Tru-Test EZ flat finish

Green
Pratt & Lambert low-luster finish
Kelly-Moore flat finish
Pratt & Lambert flat finish

Yellow
Pratt & Lambert low-luster finish
Pittsburgh Manor Hall low-luster finish

Pratt & Lambert flat finish
Sherwin-Williams flat finish

Gold
Glidden Flat flat finish
Glidden Eggshell low-luster finish

Guide to the Ratings

Listed alphabetically. Properties common to a brand apply to all colors. Dashes in columns mean a suitable color wasn't available.

Brand and model. The paints we tested are at or near the top of each line.

Gloss. As determined by CU; gloss descriptions on the label are often unreliable. Flat (**F**) is the lowest on the scale, followed by low-luster (**LL**), which looks flat when viewed head-on but shows a hint of sheen from an angle. Eggshell (**EG**) is a little glossier than flat at any angle.

Spatter. Paints that scored well in this category aren't likely to spin off onto you or other surfaces while you're rolling them on.

Scrubbing. If the wall you're painting will need frequent cleaning, choose a paint tough enough to take it. In this test, we counted the number of cycles a scrubbing machine needed to remove a standard thickness of paint. The best paints should stand up to repeated cleaning without rubbing off.

Stain removal. We mixed up a concoction of grease and soot, smeared it on the painted surfaces, and let it sit overnight. Then we attempted to remove it with a sponge dampened with a popular all-purpose spray cleaner. The paints that did best would be the first choice for walls that are in frequent contact with dirty fingers.

Blocking. Blocking is what happens when a painted surface sticks to something, even after the paint is dry. It's especially likely at higher-than-average temperatures (such as in direct sunlight) and with a surface sheen other than flat. Blocking is less important on walls than on windowsills or shelves.

One-coat hiding. The numbers in these columns correspond to the darkest stripe on our hiding chart that one coat would completely cover; 1 is the lightest stripe and 6 is the darkest. A paint that hid better covered a higher numbered stripe. In our tests, we use a fresh roller for each sample and apply the paint at its natural spreading rate. That's roughly 650 square feet per gallon, or considerably more than the typical "recommended" rate of 400 to 450 square feet per gallon listed on many paint labels.

Two-coat hiding. Any paint that scored low after two coats would be a poor choice to cover any color. If you were changing from a dark color to a light one, such a product might require three or more coats.

Fading. The better a paint scored here, the less chance you'd see color differences when you move a picture or rearrange the furniture. Paints with a low score here can fade even in indirect sunlight.

Ratings

Better ← ● ◐ ○ ◑ ● → Worse

Interior latex wall paints

As published in a **September 1988** report.

Properties common to a brand

Brand and model	Gloss	Spatter	Scrubbing	Stain Removal	Blocking
Benjamin Moore Regal Aquavelvet	LL	○	○	◉	○
Benjamin Moore Regal Wall Satin	F	◉	○	○	◉
Devoe Regency House	LL	◐	◐	◉	◐
Devoe Wonder-Tones	F	◉	◐	○	◉
Dutch Boy Dirt Fighter Flat	F	◐	●	○	◉
Dutch Boy Dirt Fighter Satin	EG	◐	◐	◉	○
Dutch Boy Super Kem-Tone	F	◐	●	◐	◉
Dutch Boy (K-Mart) The Fresh Look Flat	F	◐	●	○	◐
Dutch Boy (K-Mart) The Fresh Look Satin	EG	○	◐	◉	◐
Fuller-O'Brien Liquid Velvet	F	◉	○	○	◉
Glidden Spred Satin	F	◐	◐	○	◉
Glidden Spred Ultra Eggshell	LL	◐	◉	◉	●
Glidden Spred Ultra Flat	F	◉	◐	◐	◉
Kelly-Moore Acry-Plex	F	○	◉	◉	◉
Kelly-Moore Sat-N-Sheen	LL	○	◉	◉	●
Lucite Wall Paint	F	◐	●	◐	◉
Magicolor Satin Plus	F	◉	◐	○	◉
Pittsburgh Manor Hall	LL	○	◐	◉	◉
Pittsburgh Wallhide	F	◐	○	○	◉
Pratt & Lambert Accolade	LL	◉	◉	◉	◐
Pratt & Lambert Vapex	F	◉	○	○	◉
Sears Easy Living Flat 9400 [1]	F	◉	●	○	◉
Sherwin-Williams Classic 99 Flat	F	◐	○	○	◉
Sherwin-Williams Classic 99 Matte Flat	F	◐	◐	◐	◉
Sherwin-Williams Superpaint	F	○	◐	◐	◉
Tru-Test Supreme E-Z Kare EZ	LL	◉	◉	◉	●
Tru-Test Supreme E-Z Kare EZF	F	◉	○	◐	◉
Valspar Our Best Quality	F	◐	○	○	◉
Valspar Premium	LL	○	◐	◉	◉

[1] According to the manufacturer, this paint is being reformulated.

Properties specific to the individual color

Brand and model	Color Name (WHITE)	One-coat hiding	Two-coat hiding	Fading	Color Name (GOLD)	One-coat hiding	Two-coat hiding	Fading
Benjamin Moore Regal Aquavelvet	White	2	5	●	Base 3 GB-27	1	3	●
Benjamin Moore Regal Wall Satin	Decorators White	3	6	●	Golden Glow	4	6	◒
Devoe Regency House	High Hide White	2	6	●	Swirl	4	6	●
Devoe Wonder-Tones	White	1	4	●	Swirl	2	6	●
Dutch Boy Dirt Fighter Flat	White	2	6	●	Champagne	2	6	●
Dutch Boy Dirt Fighter Satin	White	2	5	●	Champagne	2	6	●
Dutch Boy Super Kem-Tone	White	1	3	●	Wheat Grain	2	6	●
Dutch Boy (K-Mart) The Fresh Look Flat	White White	1	4	●	Gold 2806	2	6	◉
Dutch Boy (K-Mart) The Fresh Look Satin	White White	1	4	●	Gold 2806	2	6	●
Fuller-O'Brien Liquid Velvet	White	1	4	●	Perfect Gold	2	6	●
Glidden Spred Satin	White-High Hiding	1	3	●	Medallion Gold	2	6	●
Glidden Spred Ultra Eggshell	High Hiding White	1	4	●	Nasturtium	6	6	●
Glidden Spred Ultra Flat	High Hiding White	1	4	●	Nasturtium	6	6	●
Kelly-Moore Acry-Plex	White	2	5	●	Gold I-13-2	5	6	●
Kelly-Moore Sat-N-Sheen	White	1	3	●	Buckskin	4	6	●
Lucite Wall Paint	White	3	6	●	—	—	—	—
Magicolor Satin Plus	Nonyellowing White	1	4	●	Spice Beige	3	6	●
Pittsburgh Manor Hall	White P-89-6	2	6	●	Bamboo Shoot	5	6	●
Pittsburgh Wallhide	White 80-45	1	3	●	Bamboo Shoot	2	6	●
Pratt & Lambert Accolade	One Coat White	2	6	●	Wind Song	4	6	●
Pratt & Lambert Vapex	One Coat White	2	6	●	Wind Song	5	6	●
Sears Easy Living Flat 9400 [1]	Nonyellowing White	2	6	●	Honey Fudge	2	6	◒
Sherwin-Williams Classic 99 Flat	White	1	4	●	Gold 12-20	3	6	●
Sherwin-Williams Classic 99 Matte Flat	White	2	6	●	—	—	—	—
Sherwin-Williams Superpaint	White	2	6	●	Gold 12-20	6	6	●
Tru-Test Supreme E-Z Kare EZ	White	2	6	●	Sunny Mesa	3	6	●
Tru-Test Supreme E-Z Kare EZF	White	2	6	●	—	—	—	—
Valspar Our Best Quality	White	3	6	●	Glori	4	6	●
Valspar Premium	White	2	5	●	Glori	2	6	●

[1] According to the manufacturer, this paint is being reformulated.

	PINK						GREEN			
Brand and model	Color Name	One-coat hiding	Two-coat hiding	Fading		Color Name	One-coat hiding	Two-coat hiding	Fading	
Benjamin Moore Regal Aquavelvet	Pink PR-36	1	3	◒		Green GR-103	2	5	◒	
Benjamin Moore Regal Wall Satin	Heathermist	2	5	◒		Green Whisper	4	6	◒	
Devoe Regency House	Party Peach	2	5	◒		Pistachio	3	6	◒	
Devoe Wonder-Tones	Party Peach	1	4	◒		Pistachio	2	6	◒	
Dutch Boy Dirt Fighter Flat	—	—	—	—		Mint Frost	2	6	◒	
Dutch Boy Dirt Fighter Satin	—	—	—	—		Mint Frost	3	6	◒	
Dutch Boy Super Kem-Tone	—	—	—	—		Orient Green	2	5	◓	
Dutch Boy (K-Mart) The Fresh Look Flat	Pink 1747	1	3	◒		Green 2057	3	6	◒	
Dutch Boy (K-Mart) The Fresh Look Satin	Pink 1747	1	4	◒		Green 2057	3	6	◒	
Fuller-O'Brien Liquid Velvet	Pink Puff	1	3	◒		Shadypoint	4	6	◒	
Glidden Spred Satin	Sweet Clover	1	3	◒		Forever Green	3	6	●	
Glidden Spred Ultra Eggshell	Lotus	2	4	◒		Frozen Pond	4	6	◒	
Glidden Spred Ultra Flat	Lotus	1	3	◒		Frozen Pond	3	6	◒	
Kelly-Moore Acry-Plex	Pink	2	5	◒		Green	5	6	◒	
Kelly-Moore Sat-N-Sheen	Pink	2	5	◒		Green	4	6	◒	
Lucite Wall Paint	Rose Pearl	2	5	◓		Spring Green	2	6	◓	
Magicolor Satin Plus	—	—	—	—		Mint Cooler	2	5	◓	
Pittsburgh Manor Hall	Slick Candy	3	6	◒		Frosted Mint	6	6	◒	
Pittsburgh Wallhide	Stick Candy	2	5	◒		Frosted Mint	4	6	◓	
Pratt & Lambert Accolade	Rose Mist	2	5	◒		Cool Eve	6	6	◒	
Pratt & Lambert Vapex	Rose Mist	2	6	◒		Cool Eve	6	6	◒	
Sears Easy Living Flat 9400 ①	Pink Carnation	2	6	◒		Jungle Moss Light	2	5	◓	
Sherwin-Williams Classic 99 Flat	Pink 3-3	2	6	◒		Green 24-18	2	6	◒	
Sherwin-Williams Classic 99 Matte Flat	Pink 3-3	4	6	◒		Green 24-18	5	6	◒	
Sherwin-Williams Superpaint	Pink 3-3	3	6	◒		Green 24-18	5	6	◒	
Tru-Test Supreme E-Z Kare EZ	Rose Quartz	2	6	◒		—	—	—	—	
Tru-Test Supreme E-Z Kare EZF	Rose Quartz	3	6	◒		Celery	4	6	◒	
Valspar Our Best Quality	Blush	2	6	○		Green Haze	4	6	◒	
Valspar Premium	Blush	2	6	◓		Green Haze	4	6	◒	

① According to the manufacturer, this paint is being reformulated.

Properties specific to the individual color

Brand and model	BLUE				YELLOW			
	Color Name	One-coat hiding	Two-coat hiding	Fading	Color Name	One-coat hiding	Two-coat hiding	Fading
Benjamin Moore Regal Aquavelvet	Glacier AJ-84	2	6	●	Base 2 YL-28	1	4	●
Benjamin Moore Regal Wall Satin	Country Blue	6	6	●	Chrysanthemum	2	6	○
Devoe Regency House	Blue Magic	6	6	●	Floral Yellow	1	4	○
Devoe Wonder-Tones	Blue Magic	6	6	●	Floral Yellow	1	2	○
Dutch Boy Dirt Fighter Flat	Crystal Blue	3	6	●	Sunlight	1	4	◑
Dutch Boy Dirt Fighter Satin	Crystal Blue	4	6	●	Sunlight	1	2	○
Dutch Boy Super Kem-Tone	Hazy Blue	3	6	●	Lemon Yellow	2	5	◓
Dutch Boy (K-Mart) The Fresh Look Flat	Blue 3060	3	6	●	Yellow 2816	1	4	○
Dutch Boy (K-Mart) The Fresh Look Satin	Blue 3060	4	6	●	Yellow 2816	2	4	◑
Fuller-O'Brien Liquid Velvet	Heidi	4	6	●	Marguerite	1	2	●
Glidden Spred Satin	Biscayne Blue	2	6	●	Gin Fizz	1	3	⬤
Glidden Spred Ultra Eggshell	Opaline	4	6	●	May Yellow	1	4	○
Glidden Spred Ultra Flat	Opaline	4	6	●	May Yellow	1	4	○
Kelly-Moore Acry-Plex	Blue	5	6	●	Yellow H-22-1	1	4	●
Kelly-Moore Sat-N-Sheen	Baby Blue	4	6	●	Maize	1	3	○
Lucite Wall Paint	Dove Blue	3	6	●	—	—	—	—
Magicolor Satin Plus	Blue Horizon	3	6	●	Lemon Ice	2	6	○
Pittsburgh Manor Hall	Blue Bell	6	6	●	Candlelight	3	6	●
Pittsburgh Wallhide	Blue Bell	3	6	●	Candlelight	2	5	◑
Pratt & Lambert Accolade	Azure Foam	5	6	●	Celestial Yellow	2	6	●
Pratt & Lambert Vapex	Azure Foam	4	6	●	Celestial Yellow	2	6	◑
Sears Easy Living Flat 9400 [1]	Federal State Light	3	6	●	Jasmine	3	6	◑
Sherwin-Williams Classic 99 Flat	Blue 31-18	3	6	●	Yellow 19-10	2	6	◑
Sherwin-Williams Classic 99 Matte Flat	Blue 31-18	6	6	●	Yellow 19-10	2	6	○
Sherwin-Williams Superpaint	Blue 31-18	6	6	●	Yellow 19-10	3	6	◑
Tru-Test Supreme E-Z Kare EZ	Skyline Blue	5	6	●	—	—	—	—
Tru-Test Supreme E-Z Kare EZF	Skyline Blue	4	6	●	Lemon Chiffon	2	4	●
Valspar Our Best Quality	Sky	6	6	●	Pond Lily	3	6	◑
Valspar Premium	Sky	3	6	●	Pond Lily	2	6	◑

[1] According to the manufacturer, this paint is being reformulated.

HIGH-GLOSS ENAMELS

If you're looking for shine and durability in a high-gloss paint, there's not much point in buying a latex product. It will be a bit easier to apply than an alkyd paint, but that's about all there is to recommend it. The latex probably won't shine as much, and it won't stand up to abuse.

Check the color heading in the Ratings. Look for products that scored well in hiding ability and resistance to color change. Check the columns of brand-related findings to see which products offer the best combination of other properties for which you are looking.

Guide to the Ratings

Listed in alphabetical order. Select a brand that performed well in the chosen color and in the brand-related properties.

Brand and model. Most are oil-based (alkyd) paints. Water-based (latex) products are footnoted.

Brand-consistent properties. Judgments in these columns apply to all tested colors in maker's line.

Painting ease. Latex paints generally spread more easily than alkyds.

Leveling. Brush marks should quickly disappear for a smooth finish.

Resistance to sagging. Some paints sag as they are applied, causing waves in the dried paint. A second coat won't hide the waves. The problem is worse with alkyd paints.

Hardness. One reason for using high-gloss paints is their reputed ability to stand up to wear and tear.

Water resistance. Latex paints are more sensitive to water than alkyds are—an important consideration for bathrooms, kitchens, and surfaces that will be washed often.

Rust resistance. Some of the oil-based paints say they deter rust. But none will do a good job at that unless you use a primer first. Without a primer, the oil-based paints were the better rust-resisters.

Resistance to sticking. A paint film can remain slightly sticky long after it is dry to the touch. Paints that scored well here would be the choice for shelves or window frames.

Color-sensitive properties. Performance variations are related to the type and quantity of pigment in a paint. All tested paints were factory-mixed colors. A dash means a color was not available. **Hiding** ability depends on the amount of pigment in a paint. The best paints should cover almost any previous color in just one coat. Those in the middle range will probably require two coats. The worst—most yellows—may require more than two coats, unless you're painting over the same color. **Fading** shows how much the color changes in bright sunlight. Whites that are sheltered from ultraviolet light can yellow.

Ratings

	Excellent	Very good	Good	Fair	Poor
	◖	◑	○	◐	●

High-gloss enamel paints

As published in a **March 1988** report.

Columns under **Brand-consistent properties**: Painting ease, Leveling, Sagging, Hardness, Water, Rust, Sticking, Hiding, Fading.

Columns under **Color-sensitive properties**: WHITE (Hiding, Fading), RED (Hiding, Fading), BLUE (Hiding, Fading), GREEN (Hiding, Fading), YELLOW (Hiding, Fading), BLACK, Comments.

Brand and model	Painting ease	Leveling	Sagging	Hardness	Water	Rust	Sticking	Hiding	Fading	White Hiding	White Fading	Red Hiding	Red Fading	Blue Hiding	Blue Fading	Green Hiding	Green Fading	Yellow Hiding	Yellow Fading	Comments
Benjamin Moore Impervo High Gloss Enamel	◑	◑	◑	◑	●	○	○	●	○	●	◑	◑	○	●	○	●	○	●	●	E
Benjamin Moore Impervex Latex High Gloss Enamel [1]	●	◑	◑	◑	●	●	●	●	●	◑	◑	◑	◑	◑	◑	◑	◑	●	●	C
Flecto Ferrothane Plastic Finish	◑	◑	◑	◑	●	○	●	●	●	●	●	●	○	●	◑	●	●	●	●	D
Flecto Varathane Colors in Plastic	◑	◑	◑	●	◑	◑	○	●	◑	●	●	●	◑	◑	◑	●	◑	●	●	D
Fuller-O'Brien Versaflex Alkyd Heavy Duty Enamel	◑	◑	◑	◑	◑	●	○	◑	●	◑	◑	◑	◑	◑	◑	●	●	●	●	—
Glidden Rustmaster Oil Enamel	◑	◑	◑	◑	◑	●	○	◑	●	◑	●	◑	◑	◑	●	●	●	●	●	D
Glidden Latex Gloss Enamel Series 1300 [1]	●	◑	◑	◑	●	●	○	◑	●	◑	●	◑	●	◑	●	●	●	●	●	B,D
Krylon Rust Magic Interior/Exterior Enamel	◑	◑	◑	◑	◑	◑	●	●	◑	◑	◑	●	●	◑	◑	●	●	●	●	D
Pittsburgh Interior/Exterior Enamel Gloss Oil Series 54	◑	◑	◑	◑	●	○	●	○	●	●	◑	◑	○	◑	◑	●	●	●	●	—
Red Devil Gloss Polyurethane Oil Enamel	◑	○	○	○	◑	●	●	●	○	◑	●	○	◑	●	●	◑	●	●	●	—
Rust-Oleum Stops Rust	●	◑	○	○	◑	◑	●	◑	●	◑	●	●	◑	◑	◑	◑	◑	●	●	—
Rustreat Oil Gloss Enamel	○	◑	◑	○	◑	◑	◑	○	◑	●	●	●	●	●	◑	●	○	●	●	B,E
Sears Best High Gloss Anti-Rust Oil Enamel Series 688	○	◑	◑	◑	●	○	○	●	●	●	●	●	○	◑	●	○	●	●	●	A
Sherwin Williams All Surface Enamel	◑	◑	◑	◑	◑	●	○	●	◑	◑	●	◑	◑	◑	●	○	◑	●	●	—
Tru-Test Supreme X-O Rust	◑	◑	◑	◑	●	○	●	○	●	◑	●	●	◑	●	◑	○	●	●	●	—
Tru-Test Supreme Latex Gloss Enamel Series LE [1]	●	◑	◑	◑	●	●	◑	◑	●	◑	●	◑	●	◑	◑	○	●	●	●	C
United Coatings Latex High Gloss Enamel Series 12 [1]	●	◑	●	◑	●	●	●	○	●	●	●	◑	●	◑	●	○	●	●	●	B,F
United Coatings Finest Rust Control Enamel	◑	◑	◑	◑	●	○	○	◑	●	—	—	—	—	—	—	—	—	●	●	—
Valspar Heavy Duty Alkyd Gloss Enamel Series 12	◑	○	○	○	◑	○	●	◑	●	●	◑	●	◑	◑	◑	◑	●	●	●	—

[1] Latex (water-based) paint.

Key to comments
A - Slower drying than most.
B - Less glossy than most.
C - Glossier than most latexes.
D - Yellow is more lemon than most.
E - White was less flexible than most.
F - Did not adhere as well as most; follow preparation instructions carefully.

Tool Glossary

This section is a reference for the basic tools mentioned in the text. There are, of course, many other tools that might be useful. Their omission from this list should not necessarily exclude them from your own personal arsenal of tools. We've excluded a few tools that are specific to a particular task and were described at that point in the text—the electric floor sander, discussed in connection with refinishing floors, is an example. All the tools of general use specifically mentioned in the text are in the following list, cross-referenced where necessary to produce more logical groupings.

Adjustable metal jack posts. *See* Jacks.
Ballpeen hammer. *See* Hammers.
Bits. *See* Electric drill.
Brick hammer. *See* Hammers.
Brick jointer. *See* Striking tool.
Brick set. *See* Striking tool.
Cape chisel. *See* Chisels: cold.
Carpenter's framing square. *See* Framing square.
Caulking gun. With the use of this simple and inexpensive tool, you can apply a wide variety of different sealing compounds and adhesives sold in cartridge form, designed just to fit this gun. The tip of the plastic nozzle on the cartridge may be cut to allow different amounts to be squeezed out when the gun's trigger is pulled. The cartridge is placed in

the gun and held in place by a notched plunger. When its handle points down, the plunger forces the contents out every time the trigger is squeezed. When its handle points up, the plunger releases its pressure and can be pulled back out of the cartridge, allowing it to be removed if necessary.

Chalk line. The chalk line is indispensable for marking long, straight lines (see Figure 65). The actual line may be 50 to 100 feet long, and is kept in a metal case that is filled with powdered chalk of various colors (usually blue). The end of the line is fitted with a ring or hook by which the line may be drawn out of the case and secured in place. After the line is stretched taut between two points, it is raised an inch or so from the surface and allowed to snap back, thus producing a straight line of chalk. The line must be raised exactly perpendicularly from the surface to be marked, but in all other ways it is an extremely simple tool to use. You have to rewind the line carefully back into the case after use and occasionally replenish the chalk supply.

Chisels. There is a chisel made for every conceivable purpose, but in this book we are concerned only with two broad types: wood chisels and cold chisels (see Figure 66).

Wood chisels. The most useful form of wood chisel for the work required in this book is the common plastic-handled type found in every hardware store. The chisel is usually furnished with an impact-resistant handle so that it may be hit with a hammer, while the blade may be of various sizes, from $1/8$ to 2 inches wide. Since it is a cutting tool, it should be kept sharp and its edge protected from coming into contact with other metal objects.

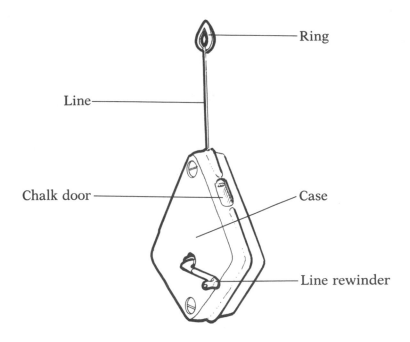

Figure 65   Chalk line

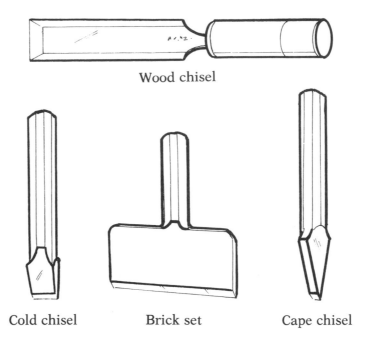

Wood chisel

Cold chisel   Brick set   Cape chisel

Figure 66 Chisels

Resist the temptation to use it as a screwdriver or as a pry bar. It is an excellent tool with which to break seams between adjoining pieces of wood.

Cold chisels. Designed to be struck directly by hammers, this type of chisel is made without a handle (although sometimes the grip portion may be sheathed with a thin layer of plastic) and is intended for hard work such as cutting metal and masonry. For general use, a ½-inch cold chisel is most useful, and should be kept filed sharp. Occasionally, types with broader blades may also be useful. When the end begins to mushroom out from being repeatedly struck, it should be filed smooth to avoid forming dangerous metal splinters. For masonry use there are a number of specialized shapes, of which the following are mentioned in the text:

Brick set. This is a broad-bladed cold chisel designed for cutting brick and trimming other kinds of masonry.

Cape chisel. The cape chisel is a short, narrow, and pointed cold chisel, ideally suited for working in crevices, such as when chipping out decayed mortar joints.

Plugging chisel. Used for similar purposes as the cape chisel, the plugging chisel has a broader blade and is consequently more effective in longer joints—both for removing old mortar and tamping in caulking materials, such as oakum and lead wool, around cracks and wood blocks set into old masonry walls.

Circular saw. Made in many sizes and with a wide variety of special-purpose blades, the circular power saw can be dangerous if not used with care. A saw that will accept a blade with a diameter of 7½ inches is the most useful for home use, and a combination blade will accomplish most tasks for which a circular saw is recommended in this book. Keep your saw clean and free of any sawdust that might impede the blade guard's proper action. Always wear protective eyewear, and never take your eyes off the blade until it has stopped moving. Roll long sleeves up and never wear a watch or bracelet when using a circular saw. Better models are equipped with easy ways to adjust the depth of cut as well as the angle of cut, and are also provided with various accessories such as fences and gauges that facilitate many otherwise tricky jobs.

Clamps. For home use, several medium-sized C-clamps will prove very useful. The better kinds have freely swiveling heads that allow the clamp to be used at different angles. That also makes tightening easier; the head does not move when the screw is tightened. Smaller, spring-loaded, clothes-pin type clamps are also invaluable, especially for light-weight holding jobs such as keeping wallpaper flat on the table as it is being pasted.

Cold chisel. *See* Chisels: cold.

Coping saw. *See* Saws.

Dividers. A pair of common dividers, preferably with legs that may be secured at any given angle, are invaluable not only for measuring off various distances but also for scribing (drawing a pattern on) the edges of boards that must be sawn irregularly to fit against less-than-perfect floors or walls. A divider looks like an oversized school compass but with two pointed legs instead of one pointed leg and a pencil.

Drill. *See* Electric drill.

Earplugs. Continuous use of noisy power tools can damage hearing, so it is only sensible to use some form of sound muffler. This can range from simple insertable earplugs to ear guards worn like headphones over the ears and rated for specific decibel levels.

Electric drill. Of all the ways invented to make holes, including bows, hand drills, braces, and breast drills, the electric drill is surely the most convenient for the home repairer. Of the various sizes available, the one that accepts a ⅜-inch bit is the most versatile. Reversible and variable-speed models are also worthwhile, even for the simplest projects. Cordless rechargeable drills, although not very powerful, are very good for light drilling and for fastening screws. Equipped with such a tool and a selection of drill bits, you can accomplish a whole range of drilling jobs easily and accurately.

Drill bits. These should be chosen for the specific job. Small sets may be purchased cheaply enough, but it may be preferable to buy individual high-quality bits as needed. Note that different bits are designed for wood and metal—the angle of the tip is ground differently and designed to be used at different speeds (see Figure 67).

Masonry bits. There is no mistaking a wood or metal bit for a masonry bit, and these should be bought separately as the occasion demands. The tip is usually made of specially hardened metal and the flutes are differently designed.

Other accessories. A whole range of accessories may be used in an electric drill, ranging from screwdriver bits (which are best used in variable-speed, reversible drills) to wire brushes, which are useful for cleaning masonry, and lamb's-wool polishing bonnets.

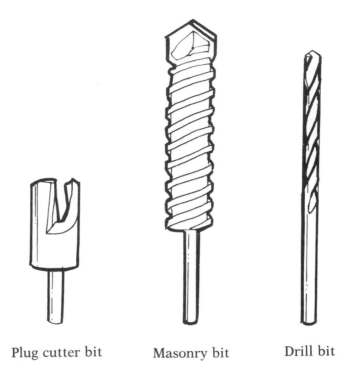

Plug cutter bit   Masonry bit   Drill bit

Figure 67 Electric drill bits

Plug cutter bits. Just one of various specialized bits, the plug cutter bit will quickly produce wood plugs from almost any wood. Although such bits can be used in an electric drill held by hand, the use of a drill stand, in which the drill is clamped, makes their use easier and more accurate. Plugs are used, for example, to cover the heads of screws in cabinetry or floors.

Electric jackhammer. *See* Hammers.

Files. Files are made in a confusing array of types and sizes, but for home use much can be done with a 10-inch, half-round, medium-coarse file known as a "bastard cut." For quicker removal of material choose a half-round wood rasp, which is a related tool that has much bigger "teeth." Since both of these are cutting tools, they should not be allowed to become dulled by contact with other metal objects. Furthermore, although sold separately, special file handles should always be fitted to any file you use; the risk of puncture wounds is great when you use a file without a handle.

Float. A float is a fairly large rectangular wooden trowel designed for smoothing freshly poured concrete. Some floats are fitted with rubber bases that facilitate the work. Test the handle—it must be comfortable if it is to be used for long periods of time.

Framing square. Also known as a carpenter's square or a carpenter's framing square, this tool is invaluable for framing wooden houses since it is calibrated with all kinds of measurements designed to produce the correct cutting angles for the studs, rafters, and other framing members. Apart from this fundamental use it is also very useful as a quick right-angle reference and straightedge (see Figure 68). The framing square usually has arms 24 and 18 inches long, arranged at right angles to each other, with the shorter arm also being somewhat narrower. Lighter models made of aluminum are less tiring to use, but they can more easily be bent out of square.

Glass cutter. Cutting glass is relatively easy if you use a sharp glass cutter. No more than a short handle with a hardened wheel mounted at one end, it is used with a straightedge to score a line on the surface of the glass, which is then snapped along this line. Should any small nubs of glass remain, the head of the glass cutter has notches cut in one side that may be used to break these off. The other end is usually formed into a small ball with which to tap the underside of the scored line, which makes the break more easily snapped.

Gloves. When using caustic or acid substances, always wear rubber gloves. When working with masonry, especially when mixing mortar or cement, wear stout canvas or leather gloves to protect your hands from the abrasiveness of the material and from the lime used in the various mixes.

Hammers. Of all the specialized hammers that are available, the most generally useful is a 16-ounce claw hammer with a crowned (slightly rounded) face. The crowned face enables nails to be driven somewhat below the surface without denting the wood, but that requires a certain amount of practice. It is generally better to use a nailset when nails must be countersunk. The claws should be sharp enough to grip small nails, but large nails should be removed with a pry bar, not with a hammer, even though a hammer has a convenient claw that may tempt you to use it. Removing nails is not really the hammer's job, and good hammers, with well-balanced hickory handles that give a certain resilience to the tool, can be damaged if you pull large nails with them. Of the many special-purpose hammers that exist, the following are mentioned in the text (see Figure 69):

Ballpeen hammer. This hammer is specially hardened so that it may be used on metal. Although its face may be flat, the edges of the face are generally beveled to reduce chipping. The purpose of the ball is to flatten rivets, but other varieties of peened hammers, such as cross-peened hammers, are equally suitable for use with metal.

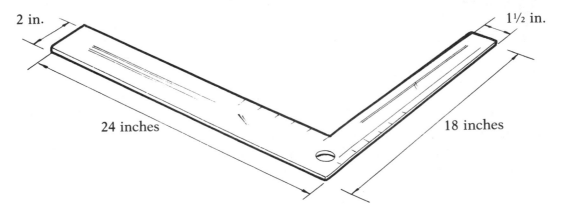

Figure 68   Framing square

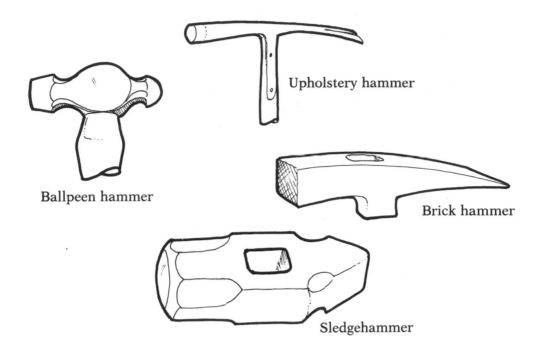

Ballpeen hammer

Upholstery hammer

Brick hammer

Sledgehammer

Figure 69   Hammers

Brick hammer. A brick hammer has a square face and curved end (in place of a claw) for trimming and cutting brick.

Electric jackhammer. There is a whole family of electric jackhammers ranging from very small to very large. Used for making holes in tarmac or breaking up concrete, it is usually best to rent one when needed. Choose the size appropriate to the job and your strength. The alternative is to use a pick—which makes for hard work.

Sledgehammer. Any large hammer with a long handle and, usually, an octagonally shaped head weighing 10 pounds or more is known as a sledgehammer. It is used for driving stakes or breaking concrete.

Soft-headed steel-faced hammer. These hammers are usually made with 5-pound heads and are ideal for use with cold chisels, since there is less danger of the hammer's face chipping when struck on metal than when using a regular claw hammer.

2½-pound hammer. A soft-headed steel-faced hammer that is made in this specific weight. It is frequently more convenient to use than heavier 5-pound hammers and sledgehammers.

Upholstery hammer. Very closely related to the tack hammer, with which it may be easily interchanged, the upholstery hammer is characterized by a long curving head and a lightweight handle, which make it ideal for driving small nails and tacks. The end opposite the face is frequently magnetized so that small nails may be more easily driven without recourse to pliers to hold them.

Handsaw. *See* Saws.

Hoe. A long-handled garden hoe may be used for mixing mortar, but a masonry hoe is stouter and better adapted to the job.

House jacks. *See* Jacks.

Jacks. Two kinds are mentioned in the text for use in raising parts of the house's structure:

Adjustable metal jack post. This is primarily an adjustable post that may be adjusted while carrying weight—thus qualifying it as a jack. Known better by various brand names, these posts may be left in place after having been used to elevate part of the structure. The several parts telescope out of one another and are then held in place with steel pins. A screw may be turned to increase the height by fractions of an inch at a time. The flanges provided at top and bottom of the post should always be securely nailed to prevent unexpected movement of the post.

House jack. A house jack or screw jack differs from a conventional lever jack by being raised or lowered gradually in a smooth and continuous motion that allows very small incremental adjustments to be made. These are crucial to maintaining the integrity of a house's structure.

Joint knife. *See* Knives.

Knives. The following varieties are mentioned in the text (see Figure 70):

Joint knife. The joint knife is an especially broad-bladed spackling knife, often as wide as 10 inches or more, used in the final smoothing of joint compound (hence its name) on gypsum-board joints. Keeping a joint knife clean is of utmost importance—even the smallest amount of joint compound allowed to dry on its edge will ruin the smooth finish expected of it.

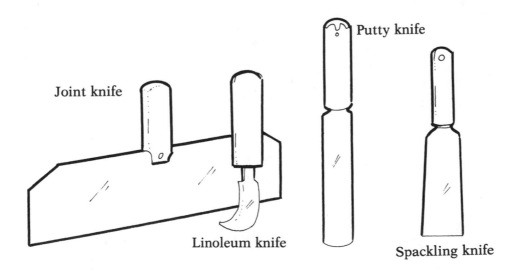

Figure 70   Knives

Linoleum knife. Linoleum is no longer manufactured in the United States, but the knives developed for cutting it—characterized by short, hooked blades—are still used for cutting similar substances, such as sheet vinyl.

Putty knife. A putty knife is a narrow knife with a pliable blade designed for smoothing putty. Putty, which contained lead and linseed oil, has been largely replaced by newer glazing compounds. A straight-edged putty knife may be used interchangeably with a narrow spackling knife.

Razor knife. There are different designs of razor knives. Some hold only specially designed blades; others are made to accept common single-edged razor blades.

Spackling knives. Spackling knives are used for applying and smoothing joint compound and spackling paste. There are many sizes and shapes, among them broad-bladed, narrow-bladed, and skewed-bladed knives. Like the joint knife, they must be kept scrupulously clean.

Utility knife. This is a loose term which embraces any knife fitted with replaceable blades. The blades are disposable and extra blades may be stored inside the handle.

Level. *See* Spirit level.

Linoleum knife. *See* Knives.

Masonry bit. *See* Electric drill.

Metal jack posts. *See* Jacks.

Metal punch. Made of hardened steel to resist chipping, a metal punch is designed to be hit by a hammer. Metal punches with pointed tips that are used for marking drilling spots in metal are known as *center punches* (see Figure 71).

Metal shears. Also known as tin snips, metal shears are simply large scissors made for cutting sheet metal such as copper or aluminum flashing.

Metal straightedge. *See* Straightedge.

Figure 71   Metal punch

Miter box. A miter box is essential to making mitered joints in woodwork. It can also help you make perfectly flush butted joints. In its simplest version, it is no more than a holding and guiding device for a saw to cut a particular angle (see Figure 72). More complicated versions are fully adjustable, allowing miters of angles other than 45 degrees to be cut with accuracy. Some have mechanisms that hold the wood being cut or secure the miter box itself when it's in use.

Nailset. A nailset is used to sink nails below the wood's surface so that the resulting hole may be filled with wood filler (see Figure 73). Using a nailset also helps prevent any accidental denting of the wood, even when the nail is merely being driven flush and not countersunk. Just a few inches long, with a knurled shank to provide a secure grip, the head of the nailset is concave so that the sharp and hardened rim may bite into the softer surface of the nailhead. Four sizes are common, but the number 2 or number 3 sizes are most useful when nailing trim.

Paper shears. Paper shears are large scissors for cutting wallpaper and other wallcoverings.

Pencil compass. The common dime-store pencil compass may be used like a divider to scribe boards to be fitted to uneven surfaces.

Pliers. Pliers are a general-purpose holding and gripping tool and are made in a variety of designs with long, short, broad, and thin jaws for holding every conceivable object. Common slip-joint pliers are the cheapest and most useful, but a pair of locking-jaw pliers, which can be adjusted and locked onto different-size objects, are also extremely useful.

Plug cutter bit. *See* Electric drill.

Plugging chisel. *See* Chisels: cold.

Plumb bob. *Plumb* comes from the Latin word for lead and has come to mean a heavy weight suspended on a line (see Figure 74). The weight causes the line to hang straight

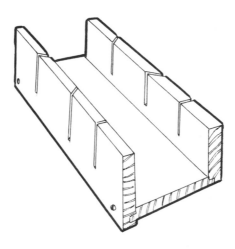

Figure 72   Miter box

Figure 73 Nailset

down, allowing you to establish or check a vertical line. Many chalk lines are made with a case the bottom of which has a point so the line-and-case may also be used as a plumb bob.

Plumb line. *See* Plumb bob.

Protective glasses or goggles. It is often absolutely essential to protect your eyes with some form of covering. Glasses with shatterproof and impact-resistant lenses are good, but sometimes more complete protection in the form of fully enclosed goggles is necessary. Always be sure that whatever you wear is appropriate for the kind of work you're doing.

Protective masks. In today's world, where so many products are made of toxic substances, any work that involves the cutting or abrading of these products should be done with the protection of the appropriate mask. Even certain natural substances, such as the dust from various woods, can be harmful if inhaled. Consequently, it is wise to choose a mask rated for the specific activity in which you will be engaged, whether it is painting or using vaporous chemicals, or creating dust particles from sawing or sanding. Sophisticated masks are actually respirators, and are equipped with replaceable cartridges that can filter out the finest of particles.

Pry bar. Pry bars may also be known as *wrecking bars* and *crowbars*, although strictly speaking these are different tools. All may be used for a variety of pulling, prying, and lifting jobs. Since they are usually equipped with a claw, they can also be used to pull nails. Made in different lengths, pry bars 18 inches long are the most convenient size for home use (see Figure 75).

Putty knife. *See* Knives.

Razor knife. *See* Knives.

Rubber gloves. *See* Gloves.

Safety goggles. *See* Protective glasses.

Safety masks. *See* Protective masks.

Saws. There are innumerable types of saw, but apart from the electric circular saw described above, the four mentioned in this book are the following:

Back saw. Usually used with a miter box, a back saw is used for finish work such as trim and moldings. The saw is rigid and has very fine teeth to ensure accurate, smooth cuts.

Coping saw. Coping saws are designed to cut intricate patterns and consequently have very narrow blades that can be rotated through a full circle. Different blades are provided for use with different materials, but all are stretched between the jaws of an open frame that may be from 5 to 18 inches deep.

Handsaw. For household use a medium-size crosscut handsaw is the most useful. The blade is quite flexible and is usually held against the surface with the handle raised somewhat. Although the teeth are designed primarily for cuts across the grain, a crosscut saw may be used for almost all small carpentry jobs around the house, providing it is kept sharp.

Sheetrock saw. This is the term commonly used to describe a short, narrow-bladed saw that is used when cutting out small areas of gypsum board or when cutting curves in this material.

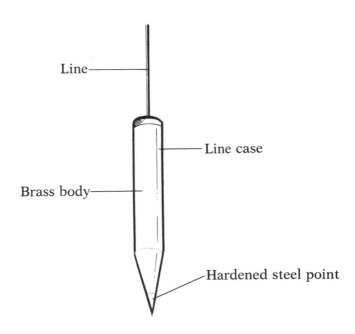

Figure 74   Plumb bob

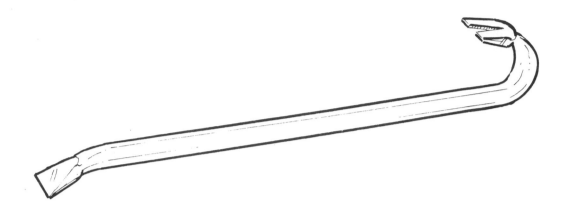

Figure 75 Pry bar

Screwdrivers. Although there are several patterns of screws, the two basic types require either a straight-bladed screwdriver or a Phillips-head screwdriver—sometimes known as the minus (−) and plus (+) types, respectively. Several sizes of each type are invariably required. It is always a good idea to buy the best, since cheap screwdrivers will not hold a sharp tip for long, and will damage the screws on which they are used.

Seam roller. A seam roller is a hard rubber or wooden roller held in the end of a short handle, which is used to press down the seams of wallpaper. It is also useful for rolling down areas that have been reglued.

Sheetrock saw. *See* Saws.

Sledgehammer. *See* Hammers.

Spackling knife. *See* Knives.

Spade. For concrete work, a long-handled spade with a rounded blade is best, rather than the short-handled garden digging spade.

Spirit level. For woodworking, a spirit level made of wood is preferable to a metal one since there is less chance of the wood marring any finish work (see Figure 76). Masons' levels are made of metal since wooden levels wear too fast when in constant contact with abrasive masonry. Both types are made in different lengths. Short 2-foot wood levels are useful for enclosed spaces, 4-foot levels are better suited for framing. Of course, the longer the level, the easier it is to obtain an accurate reading over an extended distance. About 3 inches long, a line level has hooks at each end from which it can be hung from a taut string. The string then acts as a level. Line levels are very useful for establishing level over long distances. Care should be taken in handling a level since it can lose its accuracy if the tubes holding the fluid and bubble are jarred. For this reason, some spirit levels are made with adjustable sights to compensate for accidental inaccuracies. Although commonly called levels, most spirit levels can also be used to measure vertical alignment, or plumb.

Staple gun. A beefed-up relative of the desk stapler, the staple gun is a specialized spring-loaded tool for driving staples of varying length, and, with the help of slip-on attachments, of varying shapes.

Straightedge. A straightedge may be any straight edge, but the term is usually taken to mean a long metal ruler, at least 3 feet long, and graduated in inches or centimeters. It is used not only as a measuring tool but also as a guide against which straight lines may be drawn or cut.

Striking tool. Striking tools may also be known as brick jointers and come in a variety of patterns of which the tuck pointer is but one. Their purpose is to firm and smooth the mortar in the joints between adjacent courses of bricks. This is a simple operation that can be done with a number of other objects, but striking tools are formed to leave particular shapes, from flat to convex, and being made of metal are long-lasting and easy to clean.

T square (for gypsum board). There are a number of special-purpose tools designed to make the life of the gypsum-board worker easier. Of these, the giant T square is one of the most useful. The center tongue, the stem of the *T*, is a full 4 feet long, so it will reach completely across an entire sheet of gypsum board and enable a perfect right-angled cut to be marked and made. It is also graduated and thus very useful for measuring cutouts.

Tile nippers. For trimming ceramic tile, tile nippers are an improvement over regular pliers since their jaws are broader and, being parallel, do a better job in breaking off irregular edges (see Figure 77).

Trowels. Although the trowel is a very simple and basic tool, it comes in different sizes and shapes (see Figure 78). A triangular trowel is best for most masonry work; a small triangular trowel is especially useful for pointing joints and reaching into restricted spaces. Square or rectangular trowels are used for smoothing plaster and concrete, and toothed rectangular trowels are made to apply adhesives, such as mastic, over large areas in an even layer.

Try-square. The try-square is used to lay out or measure an exact 90-degree angle. Consisting of a blade set in a stock, it must form a perfect right angle if it is to do its job. Cheap wood models may be sufficient for rough work, but if accuracy is required, machinists' squares are a better investment. Made in the same range of sizes as wood try-squares,

Figure 76 Spirit level

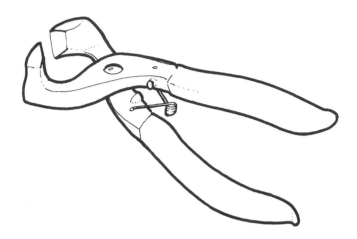

Figure 77   Tile nippers

both blade and stock are made of metal and very securely and accurately joined together.

Tuck pointer. *See* Striking tool.

2½-pound hammer. *See* Hammers.

Upholstery hammer. *See* Hammers.

Utility knife. *See* Knives.

Wire brush. Used for all manner of stripping and cleaning work, the wire brush is made with stiff carbon-steel bristles folded into a wood handle and secured with staples. Since carbon steel rusts easily, care must be taken to keep this brush dry. Buy brushes with curved handles, since these will keep your knuckles from rubbing against the work.

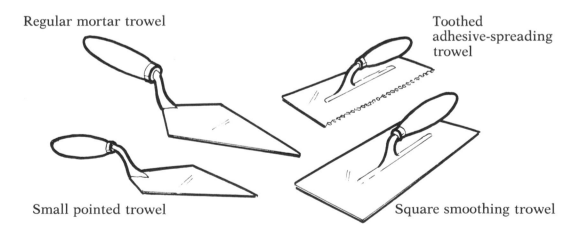

Regular mortar trowel

Toothed adhesive-spreading trowel

Small pointed trowel

Square smoothing trowel

Figure 78   Trowels

Index